Why Not Say It Clearly
A Guide to Scientific Writing

Lester S. King, M.D.

Why Not Say It Clearly
A Guide to Scientific Writing

Little, Brown and Company
Boston

Published November 1978

Copyright © 1978 by Little, Brown and Company (Inc.)

First Edition

Second Printing

Library of Congress Catalog Card No. 78-69907

ISBN 0-316-49346-5

Printed in the United States of America

SEM

Preface

In a sense this book had its inception in 1963 when I joined the editorial staff of the *Journal of the American Medical Association.* A few months after I began work, Dr. John H. Talbott, the editor, asked me to initiate a course in medical writing for medical students. After the first year Dr. Charles Roland, newly appointed senior editor, became associated with the project. We agreed that the focus of the program should change from the medical student to the physician (either in practice or in a residency). With the occasional help of teachers outside the AMA we gave numerous courses, the longest of which lasted six weeks and the shortest, one day. Most satisfactory were those workshops that lasted one week.

After a few years Dr. Roland and I embodied our experiences in a series of brief articles, first published in the *JAMA* and later collected into a small book entitled *Scientific Writing* (American Medical Association, 1968). Although Dr. Roland left the *JAMA* shortly thereafter, he and I continued our efforts in medical journalism, sometimes in collaboration, more often in separate programs.

Scientific Writing was a joint effort in which the contributions of each of us were clearly indicated. After a few years I found that the parts I had written no longer adequately expressed my ideas. Drawing on further experience and reflection I have recast the format and written an entirely new text. From the earlier volume I retained a few examples that I had already used, but the text is totally rewritten to constitute a new book.

Why Not Say It Clearly is a personal credo. It does not represent the views of any editorial board or publisher but reflects the personal standards that I have developed in more than half a century. My credo is, in brief: At all times there are certain pressing needs—for flexibility, for awareness of possible alternatives, for judgment (that, in a given context, one alternative is better than another), for conscience (that drives the writer to find the best mode of expression). It is extraordinarily difficult for a writer to make clear in his own mind what he really wants to say. Even more difficult is the task of saying it in a fashion both lucid and pleasing. With this set of underlying beliefs I point out modes of writing that hinder clear expression and I offer suggestions to help say things more clearly and agreeably.

Anyone who writes on "how to write" is vulnerable to hostile critics. In this regard I have twice had especially interesting experiences. In each instance, after I had reviewed a book that dealt with writing, I received long letters pointing out stylistic flaws in my own reviews. For reasons that are not clear to me, the letter-writers wanted to show me that what I had tried to say could be said "better." In essence, the correspondents rewrote my reviews to accord with their own stylistic preferences. One writer did point out a grammatical boo-boo I had overlooked, and for this I was grateful. Otherwise I flatly disagreed with most of the proffered changes. The alterations I considered to be stylistically undesirable. Yet to my critics I replied briefly, in words that I believe stemmed originally from H.L. Mencken, "You may be right." And I thought to myself, without putting it on paper, "But I am convinced you are wrong." I did not try to argue.

Those two incidents I mention to emphasize an important point: I do not set myself up as an authority. I merely express my own views and hope that many will agree with me. By virtue of appearing in print, what I advocate takes on a certain respectability and can serve as a demonstrable reference—an authority of sorts. But obviously many will disagree on the grounds of personal stylistic preferences. I try to make explicit some available linguistic choices and to recommend certain of those choices in preference to others. This book may serve as a reference, a guide, and perhaps an instructional manual, but not as an "authority."

The illustrative examples in this book are of two major types. Some, taken from standard works of English literature, are printed verbatim with appropriate bibliographic reference. Most of the examples, however, come from medical journals. In the excerpts, I have suppressed references and introduced minor changes to hinder identification of authorship or source. For the most part I have changed the names of diseases, altered various technical terms, and used different dates and numbers. Furthermore, all proper names mentioned in the originals have been transformed into Doe and Roe.

All of the examples are real, that is, not made up for the occasion (with the obvious exception of the one on page 44). Furthermore, I have never intensified a fault to make an example more striking. All defects occurred in the originals and I have never made a quotation worse for the sake of effect. On the contrary, if an excerpt contained more than one severe fault, I might, for better pedagogic effect, correct all but the fault under discussion. However bad the quotations in the book may seem, the original texts were at least as bad, and possibly worse.

To provide material for discussion I have relied largely on medical situations, but the problems I discuss are in no way limited to medicine. Since in essence I deal with expository prose (in contrast to narration or argument), I hope to reach various circles of readers and writers: most immediately those who engage in medical writing; then, those who deal with scientific topics, in general; and finally, those who engage in any type of exposition, regardless of subject matter.

This book owes much to my many students, whom I want to thank for having stimulated the flow of thought. I also want to thank the editorial staff of Little, Brown and Company, especially Ms. Gretchen Denton and Ms. Christine Ulwick, who rescued me from many infelicities of expression. For any that remain—and I have every expectation that critics will not be backward in pointing them out—the responsibility is entirely mine.

L. S. K.

Contents

Why Not Say It Clearly
A Guide to Scientific Writing

The Present Scene

Why Is Medical Writing Bad?

In the past 16 years as an editor of the *Journal of the American Medical Association*, I have had occasion to examine a vast number of manuscripts submitted for publication. The experience has often led me to ask myself, Why do doctors write so much? Why such a passion for putting words on paper with the hope of getting these words published? Various answers suggested themselves. When I felt kindly disposed to my fellowmen, I thought that the writers were convinced that they had made discoveries or observations that significantly advanced knowledge. Since knowledge is one of the great glories of civilization, it should be shared. To be sure, many writers may have been mistaken in this evaluation of their efforts, but at least their motives were sound. In my cynical moments I suspected more venal motives—that the driving force was the desire to amass a bibliography, win grants, gain academic promotion, or enhance professional status. In my dispassionate moments, I realized that of course motives are mixed and there are no simplistic answers.

More to the point is the equally persistent question, Why do doctors write so badly? In this regard I know that medicine is not worse than other professions, that sociologists and psychologists, philosophers and educators, as classes, write just as voluminously and just as opaquely. Why, then, do members of the learned professions write so badly? Some critics want to blame the entire educational system, starting with primary school; others, more selective, would lay the blame principally on college or perhaps on graduate school. Still others might want to involve our entire civilization and its scheme of values, with special emphasis on television. I have no desire to enter a controversy of this scope or to offer comprehensive answers or far-reaching and ultimate explanations. I want merely to study that small area where an editor makes contact with his contributors, to seek a few proximal explanations for the bad writing, and in the course of this book, to offer suggestions for achieving improvement.

Present-day medicine, with its scheme of values and rewards, exerts considerable pressure on physicians to acquire a bibliography. Those who seek any sort of academic career, whether full-time or part-time, must have a list of publications if they want to climb the academic ladder. Publish or perish is no idle imperative, for publication can lead to academic promotion; failure to publish, to academic stagnation. Of crucial importance here is the research grant that shapes academic medicine. There is an elegant circularity: A list of previous publications makes it easier to win a grant, and a grant once obtained makes it easier to get a larger bibliography. In this scramble to win grants and to get papers into print there is no comparable pressure to write well.

I have heard some research workers, who recognize this phenomenon, offer the excuse of technical specialization. Each profession, they say, has its *in-group* that has mastered a special vocabulary and special concepts that the uninitiated are not expected to understand. The members of the in-group understand each other and those outside the group do not matter. One investigator emphasized that research workers have no difficulty in understanding one another and have no need to write for general practitioners (or members of other groups) who are out of touch with the latest developments. Another physician declared that if an author has something important to say, he need not concern himself overmuch with the way he says it. A copy editor can always untie the knots in his prose, and the author himself need not bother with the niceties of skillful writing.

According to this philosophy the researcher should shut himself up in an ivory tower with a group of like-minded colleagues who speak only to each other, while the benevolent patron who pays the bills demands nothing more in return than publication. While this attitude is by no means dominant, it does exist to an uncomfortable degree. It represents an indifference to good writing. Now, if we compare the members of a learned profession who write for publication with laymen who in one or another way make their living from the written word—whether journalist, script writer, novelist, or essayist—we find an important difference. The lay writer who makes some or all of his livelihood from his pen works in a highly competitive field where style counts. How he says something is as important as what he has to say, and the presentation of ideas carries as much weight as the ideas themselves. In such a situation a lack of skill is self-destructive. Those who must write well to survive will learn to write well—or they will not survive. This necessity does not apply to medical writing or scientific communication generally. Competition exists, but of a different character.

Among professional journals authors compete for publication space, journals compete for "good" manuscripts. Authors want prompt acceptance by a prestigious journal and prompt publication; editors want well-written manuscripts that are professionally important. Editors must also have on hand a comfortable backlog of material, sufficient to ensure continuity of publication, yet not so great that the publication of new manuscripts will be unduly delayed. Authors and publishers thus live in symbiosis. The unpublished manuscript accomplishes nothing for its author, and a journal without manuscripts speedily dies. This interdependence results in some curious tensions directly relevant to writing skills.

Professional journals differ greatly in subject matter, size, degree of specialization, frequency of appearance, circulation and readership (by no means the same thing), editorial policy, and prestige. Some journals are highly specialized, others are more general; some have an international reputation, others are purely local and parochial. Prestigious journals receive large numbers of manuscripts and have a correspondingly high rejection rate that may exceed 80 percent. Authors whose manuscripts have been rejected do not lose heart or discard their brainchildren. Instead they select another journal as second choice; and if not successful here they can try still another and then another. Journals are so numerous that almost any manuscript, even if trivial and badly written, can eventually get published somewhere, although the journal may have a low circulation and little prestige.

Functions of the Editor

The journal editor controls the sluice gates to publication, but he operates under certain pressures: He must receive an adequate number of manuscripts from which to choose, and his choice must enhance the prestige of his journal and, if possible, increase its circulation and influence. In his process of choice he must regard both the content and the form of manuscripts submitted. He may call on referees to help him evaluate the merits of any given paper, but referees ordinarily pay little attention to modes of expression. The editor must decide whether a paper acceptable for the significance of its message is written with adequate clarity and grace. This editorial function is all too often neglected.

Let us take two extreme cases. Some editors, unfortunately, have little skill in writing. Indeed, judging by their own signed editorials and the literary quality of the articles they have accepted, these editors have neither the ability nor the capacity to raise the standards of medical writing. Instead, they contribute mightily to the low state of medical communication.

At the opposite extreme is the highly literate editor whose own writings are a pleasure to read and who has a fine sense of style. What can he do to raise the level of medical prose? He can impose sound standards and demand clear writing as a requirement for acceptance. This attitude, however, has two drawbacks. The editor may want to work with an author to improve the paper, but this process can prove so time-consuming that on a purely practical basis the editor may have to compromise his principles. Or, if the editor demands repeated and extensive revision, he may alienate the author, who could have had his paper published elsewhere with much less reworking. Standards that are too high can divert contributors to less fussy competing journals. Many marginal journals cannot afford to be choosy and are glad to receive contributions that have an acceptable content even if badly expressed. The world might be a better place if some marginal journals ceased to exist.

Is it quixotic to hope that editors as a class will raise their standards and insist that papers be better written? Criticism that journal editors are too lax is by no means new. More than a century ago critics deplored the repulsive quality of medical prose, as John Blake has demonstrated in an excellent survey [1]. Shortly after its founding, the American Medical Association established committees to evaluate the medical literature. The committee of 1851 criticized editors for publishing papers that reveal *not only want of rhetorical finish (a slight blemish, comparatively speaking), but of all regard to correctness or appropriateness of language. . . . An inexcusable defect in composition, for the reason that it is so easily avoided, is the commonplace, inaccurate, in short, illiterate, language suffered to find its way into our journals.*

The committee report of 1852 thought that an editor

is bound to see that his contributors do not offend the common rules of language and to refuse their papers if devoid of merit or incurably loaded with faults.

In his splendid essay Blake, without drawing conclusions, let us infer that even if present-day writing is bad, it was worse at an earlier time. I would like to add a single example to Blake's article, the lament of an editor in 1900 regarding the excess verbiage of published articles [3].

The majority of articles submitted for publication could be cut down one-half, and not a thought be eliminated in so doing. The repetition of well-known facts, padding with abstracts from text-books, and words, words, words, too often constitute the papers that appear as "original" in medical journals. And if the editor presumes to use the blue pencil in the least, the majority of authors consider it an insult.

If the fault is so clear, and has been for so long a time, why is the remedy so difficult to find?—a question easier to ask than to answer. Yet every editor should ponder the question and find an answer.

Some Problems of Publishing

New journals spring up like the proverbial mushrooms. Every month, it seems, I get announcements of this or that new periodical, with proudly displayed editorial board, a solemn statement of goals, and a list of papers scheduled for early publication. Most of these new journals flourish. Since World War II the mortality of specialty journals has markedly declined, and if we analyze the factors, the main feature will be money available for research and for hospital and medical school expansion.

The research grant has become not merely a way of life but a very source of life which, if withdrawn, would have a disastrous effect. We may think of a mighty river, like the Colorado, on which so many functions depend: If the rainfall or snowfall is insufficient, the level of the river drops and cries of anguish arise across the land. The support of research is comparable to the rain or snow that keeps the Colorado at a high level.

But research demands publication. If existing journals are not adequate for the load, new journals must be established. The new journals in turn are nourished by the flow of manuscripts until, if that flow becomes a torrent, further journals will make their appearance. Their field is usually not identical with the old, but represents rather a minor degree of specialization. Science grows by multiplication of research, of manuscripts that represent the results of research, of journals that record these results.

Apart from the research grants, certain other economic and social factors have great importance. Publishing a journal costs money, and is costing more and more every year. Where does the money come from? Excluding special subsidies there are two major sources, advertising revenue and subscriptions.

I will mention only two facets of advertising. The first is the so-called controlled circulation journal, sometimes known as a throwaway. Here the entire cost is born by the advertisements; the journal is distributed free to a select group, e.g., all licensed physicians in practice or those in particular specialties. Journals of this type exist to make a profit. To do so they must be readable, so that the advertising will be noticed. The editors of such journals must assemble reading matter sufficiently interesting that the readers will at least turn the pages and not cast the journal immediately into the wastebasket.

These journals ordinarily show some sort of specialization—one concentrating, say, on news and the latest professional advances on all fronts; another on specific economic problems affecting the readership; another, on cultural phases; another, on digesting the new literature into readily assimilable form. If skillfully done this reworking is clear and readable, often more so than the original papers. Journals that offer predigestion of material and clear presentation have a great vogue. Physicians like—and read—them but may neglect the more prestigious journals that give original publication. And advertisers patronize the journals that are read. The physician who tries to keep up may assiduously read the professional journals that have some academic stature; or he may pay lip service to these and get most of his information from the throwaways.

Subscriptions are the second major mode of financing publication. Here we must distinguish two types. Many societies, among their activities, publish a journal, the receipt of which is a benefit of membership. If virtually all membership dues go to supporting the journal, the journal can truly flourish. But if only a part of the dues goes to support it, publication costs may need to be met partly by advertising, with the various drawbacks that this can entail.

Certain journals, however, get most of their income from subscribers who are not members of the parent society. We can distinguish a captive audience, who, by virtue of their membership, receive a journal they do not necessarily read; and "free will" subscribers who pay money specially to get hold of the journal, presumably because they do want to read it. Such journals are less dependent on advertising than those that cater to a captive audience.

The amount of advertising can affect the acceptance of manuscripts. To qualify for favorable postage rates a journal must maintain a suitable balance between editorial matter and advertising or else pay higher postal rates. If the quantity of advertising is high the journal, to qualify for the favorable postage rates, must have much more editorial matter. It will welcome papers and may even be hard pressed to get enough editorial content to "cover" the ads. On the other hand, when the ads are few, a journal dependent on this form of revenue will not be able to afford so much editorial content. A small volume of advertising may thus force such a journal to reject papers that in more lush times it might deem acceptable.

We can thus glimpse a tangled sequence. A journal may want to attract advertisers. For this it must have a high readership (not the same thing as a high total circulation figure); to attract readers the papers must be readable and interesting, to increase prestige, they must have significance. The quality and reputation of a journal depend on the editor, who must evaluate the different parts of this sequence. He must decide on the audience to which he is appealing,   9 the function he is trying to serve, the degree to which advertising may affect policies, the aggressiveness with which he will search for authors to fulfill his intentions, and the emphasis he will place on quality of writing. A strong aggressive editor can markedly increase the readership (and the circulation) of his journal; a weak editor can speedily run a journal into the ground.

The editor must face squarely the problem that Franz Ingelfinger phrased so well. In an editorial in the *New England Journal of Medicine* [2] he pointed to the difference between education and information. "The informed physician has facts at his disposal, but only the educated professional understands their uses." He felt that general medical journals should have the function of educating and fostering understanding. "Such journals should be like universities, not like information booths." This is one difference between the throwaway, that provides information, and a more prestigious journal that truly educates. On the other hand, we face the recalcitrance of the medical profession who, in large part, want readily usable information rather than education and understanding. And there are also economic realities. Here is a clash of values. An editor must take a stand on one side or the other, or possibly try to arrange a compromise.

Editors, with the best will in the world, may nevertheless not have the power to produce the kind of journal they wish. Editors do not own the journals they edit. A publisher—an individual or corporation or organization—owns the publication, and the editor is an employee. The owner—for our purposes, usually a society or association, with a board of trustees—has the last word. This board may set policies to which an editor must conform. However, a really strong editor, who is also a good editor, can usually persuade a board to see things his way.

Such problems may seem far removed from the troubles of pure research, the awarding of research grants, the writing up of the results, and publication. However, research takes place not in an ivory tower but in the real world—perhaps rather distant from the marketplace, but still not so far as to be unaffected thereby. Let me give a single example.

We are now in the midst of considerable economic stringency. A short time ago, when the stringency was perhaps even greater, some journals faced a great increase in costs of production and a decline in advertising revenue. To meet this threat, by cutting costs, the publishers decreed an overall diminution in the number of pages allotted to these journals. Here was a glorious opportunity for editors to enforce high standards, to insist that papers be shorter, better written, and more intelligible. Most manuscripts—at least 19 out of 20—would be improved by substantial shortening. The editors could have demanded that manuscripts otherwise acceptable be materially reduced before final acceptance. Unfortunately, in the instances I have in mind, the editors did not adopt this belt-tightening process but, through various pressures on the publishers, secured a restoration of the cuts. A heaven-sent opportunity for improvement was ignored. Where lies the fault?

The relations between author and editor (and indirectly, publisher) rest on different and often contradictory values, each of which may give rise to special problems. Involved are the agencies that fund research, the author, the editor, the publisher, the advertisers, the reader, all of whom have different wants that may conflict. In this complex skein good writing—clear, informative, and pleasing—is only one value among many. However, I believe it should have a high priority. In this book I present my own views on what constitutes good expository writing and on the ways to achieve it.

References

1. Blake, J. B. Literary style in American medical writing. *J.A.M.A.* 216:77, 1971.
2. Ingelfinger, F. J. Purpose of the general medical journal (editorial). *N. Engl. J. Med.* 287: 1043, 1972.
3. Medical literature and medical writing (editorial). *J.A.M.A.* 35: 626, 1900.

Good and Bad Writing

"Good" versus "Bad" Writing

In this book I speak of "good writing," and "bad writing," and the notion of "better." When I apply these terms, I lay myself open to attack. "Good" and "bad" imply standards. What are the standards with respect to English prose? Who sets them? Who interprets them? What compulsion do they exert? To avert pointless controversy I will make clear at the outset the way in which I use these terms relative to writing.

Let me offer an oblique illustration, involving accuracy rather than goodness. Suppose you want to answer the question, "What is the correct time?" You look at your watch, and then, realizing that perhaps it may be slow, you check with the radio or television, the telephone service, or the chronometer in the jeweler's window. Those sources you regard as reliable standards, yet they do not precisely agree. Nevertheless, they give you a sense of good enough, of accuracy sufficient for practical purposes. If you want a still higher degree of precision, you might consult with the astronomers who read the movements of the celestial bodies, and finally you might get entangled in concepts of relativity. You will soon conclude that there is no absolute correct time, and that search for an absolute is futile.

Under ordinary circumstances you accept as the "right time" the answer given by a source you accept as the authority in that particular context. But certain contexts may raise difficulties. For example, the session of a state legislature may by law terminate on a given day at midnight. If midnight approaches when there is still a lot of unfinished business, the sergeant-at-arms may stop the clock while the legislature proceeds with its deliberations. "Midnight" arrives only when the appropriate authority officially declares that it does. If the sergeant-at-arms says that the time is 11:45 P.M., it is quite irrelevant, *in that context*, if the radio signal says 5:00 A.M..

A somewhat analogous situation is the recording of the time of death. A patient in a hospital is not officially dead until an authorized physician pronounces him dead, and there may be a significant interval between biological death and the physician's pronounce- **13** ment. Ordinarily this delay causes no great difficulty, but we can conceive of grave problems in certain cases of inheritance in which the title of property passes at the moment of death and the precise time may be crucial.

The right time, then, is what some accepted expert *says* is the right time. In some contexts we have the option of disbelieving the assertion and seeking a different answer elsewhere. But in other contexts we have no option—the person who made the assertion has irrefragable authority and if we do not believe him there is not much we can do about it.

I suggest a close parallel to assertions about good writing. Judgment depends on context, and good writing is what some expert says is good writing. But why should we accept that authority? Sometimes there is no choice. Ordinarily, for example, there is no appeal from the decision of the schoolteacher who gives a low grade to the composition of a pupil or the decision of an editor who rejects the manuscript of a would-be contributor. The teacher and the editor become authorities by virtue of their positions, and through their positions they set standards. Of course, another teacher or another editor need not agree. The pupil may go to another school and get a better grade; the author may submit his manuscript to another journal and get an acceptance. Thus we can appreciate the conflict of standards.

The judgments of the teacher or editor, which may have practical consequences for the pupil or the aspiring writer, are nevertheless subjective and relative. There is no absolute or objective measure of good writing. Often, in our desire for objectivity, we turn to books, for the printed page seems to exert a certain power over our judgment. But as we consult various dictionaries and texts on rhetoric or on usage, we learn that dictionaries may disagree with each other and that even in a given dictionary the various contributing experts may differ markedly in their opinions. Generally accepted usages change in the course of time. The "vulgar" may become accepted in "polite" society, and the acceptable may become obsolete.

When I speak of good writing, I am implying relative and not absolute standards. I am saying merely that some writings are better than others; the better ones fall into a category I call "good," and the worse ones I call "bad." The particular epithets I apply may depend on the context. When I call some writing bad, I mean that it needs improvement; when I call an example good, I do not mean that it cannot stand improvement but that its present form is quite satisfactory—within its context. My value judgments depend on comparison. If we make some appropriate change, is the result *better* than the original? And in this book I let the reader decide whether one example is better or worse than another. The reader should depend on his own judgment. All I do is try to help him make up his mind, to give him some grounds for decision, some tools for his analysis.

Developing a Critical Sense

When I give courses in writing for physicians, I start by asserting that some writing is good and some bad, and I ask the class to be the judge. To show how easy is evaluation, I suggest a simple schema: I present numerous examples which I ask the class to rate as either good or bad without further qualification.

The examples, of course, are carefully—I might say artfully— chosen, and usually the students in the class agree completely on at least 80 percent of the quotations. Some examples provoke a difference of opinion. Then I ask the truly difficult question, *Why* do you call this one good and that one bad? I ask the students to justify and defend their opinions. At this point they usually have the utmost difficulty. They offer a few indefinite criticisms, commonly applying terms like *vague, unclear, confused* to the examples judged bad and *clear, simple, interesting, informative* to those judged good. Obviously the students could not provide any really defensible grounds for their judgment. They had an intuitive reaction to the examples but no real grasp of the stylistic qualities that had induced their particular judgment.

I then try to provide an analytic framework that will help the students reach their own conclusions and justify their decisions. Naturally, I offer my own scheme of values, which they may accept or reject or modify as they please. But in presenting my own viewpoint I constantly stress the reasons behind my opinions, and if the students disagree, I insist that they formulate reasons for their own judgments. In this way I seek to inculcate a habit of critical analysis. And in the present book I use the same approach.

First, however, I want to offer concrete examples of what I mean by good and bad writing. Ten short paragraphs follow: Six of these I would place in the category of bad writing, four I consider good. In some the meaning is immediately clear, in others, obscure to the point of bafflement. Some are attractive, and please the ear when read aloud; others are repulsive. At this point I offer no analysis. I do not discuss the factors that induce the value judgments nor the ways of eliminating the faults. All that will come later. Here I wish only to make vivid the distinction between bad and good writing, between writing that sorely needs change and writing that does not.

The results of the present study suggest that in addition to the manifestation of aberrant homeostatic patterns of neurohumoral activity following the cessation of noxious stimulation, the neurotic may be further characterized by atypical autonomic responses to an increase in the level of appetitional drives.

The book is essentially a potboiler. Although abundant research has gone into it, the text is largely scissors-and-paste—excerpts taken from contemporary sources and joined together by facile prose. No real picture emerges. There is no synthesis. *The author has not digested anything. We do not get a real three-dimensional picture of the times; we do not get any insight into the medical practice of the era; we learn neither American history nor medical history; and we are not very much entertained.*

The unique old world charm in scenic country-like atmosphere with a large variety of individualistic newer as well as charming older homes of various styles on larger wooded parcels makes Lake County the favorite of a large number of Detroit executives.

For 30 years as a practicing pathologist, I have been trying to supply the answers to the perpetual question, "Why did he die?" It has become most fatiguing. The conscientious clinician very dutifully inquires after the autopsy findings. I may show him a Pandora's box full of pathology: ruptured viscera, infected and torn heart valves, thrombosed arteries, widespread cancer, or a peritoneum filled with exudate. "I know all that," says the earnest clinician with a tone registering anything from mild reproof to utter disdain, "but why did he die?" Eventually this proves very bad for my disposition.

In the current study, explicit process criteria were developed using an informal group setting, with a small but consistent group of practitioners meeting on numerous occasions over an extended time period to achieve an eventual consensus on minimal management criteria for the care of four indicator conditions.

In my opinion, part of medicine's neglect has been due not only to uncertainty concerning etiology, treatment, and prognosis but also that acceptance of homosexuality as a medical disorder alongside all other medical disorders has been unconsciously and consciously perceived by us as tantamount to being in favor of it, encouraging it, and perhaps endorsing it, thereby putting us in direct conflict with established standards of human conduct.

The laboratory director who seeks a program of quality control sometimes feels like Goldilocks at the bears' home: Some suggested programs are Too Big, others are clearly Too Small. This handbook outlines procedures and suggests combinations that can be Just Right for each individual laboratory.

The pneumonia was treated with penicillin, then changed to erythromycin.

This topic has been a matter of discussion ever since medical education dedicated itself to the goal of making the practice of medicine a scientific discipline based on the scientific principles of biology, chemistry, physics, etc. With the acceptance of this dedication, there has been continued discussion of the role of the basic sciences and there is very little more that I can say on this topic that has not already been said before by many medical educators. However, the basic issue persists, viz, that if medicine is to continue its dedication to becoming a scientific discipline it must be prepared to incorporate all scientific advances made in the basic areas of biology, chemistry, physics, etc., as they apply to man.

Dr. Doe's coronary attack had beneficial results, for during his convalescence he tape-recorded his reminiscences, giving us not so much a formal autobiography as a series of anecdotal vignettes. These, excellently edited, make up the present volume, and a delightful volume it is.

Content, Form, and Judgment

Customarily we distinguish between content and form, between what we are trying to say and the way we say it. Whether these aspects are *really* separable is, perhaps, an open question, and one to

17

which we will return in a later chapter. For the present I will treat them as separable. The questions, How can you improve what you have said? and How can you make it better? refer to form. The notion of "better" implies not only standards but also the existence of techniques for meeting those standards. Specific details of technique I will consider later. Here I will take up only some general considerations.

As an editor I may receive a manuscript that conveys an important message, yet is so badly written as to be quite unpublishable. How to get it into suitable shape? Obviously, by trying to have the author improve it. But usually an author who produces a badly written paper does not know how to do better. Offering advice and suggestions would not make any impression, nor would I make his task easier if I recommended that he read some standard books on writing—full of good advice that is usually singularly unhelpful. Thus, the books may tell him to eliminate redundancy, get rid of fancy words, avoid loose sentences, and use strong active verbs. But how will the poor author identify which words are needless, which sentences are loose, or which verbs are strong?

We have a comparable situation when we tell a person not to drive too fast or eat too much. Such exhortations involve a scale of values. How fast is too fast? How much is too much? We can easily teach a person some rules, but deciding whether a rule applies in a given case requires not memory but judgment. And judgment cannot be taught. It can only be trained or developed.

The writer who wants to improve his style must develop judgment regarding the values involved in writing. He must learn to recognize good writing and distinguish it from bad. He must then become sensitized to bad writing. He must constantly try to discriminate the good from the bad, whether he is reading a newspaper, a medical journal, a volume of history, a treatise in philosophy, or a textbook of pathology.

In bad writing, a characteristic warning flag is a difficulty in understanding what the author is trying to say. You may "stumble" as you read, so that even when you know the meaning of each individual word, you fail to grasp the sense, or you arrive at it only after several readings. Of course, we are presupposing a certain level of maturity. A child might readily understand a topic discussed in a children's encyclopedia but not the account in a college textbook, no matter how well written the latter might be. But the mature and reasonably well educated adult who cannot understand expository writing should not conclude that he is dull-witted or that the subject is too abstruse for him. He should at least entertain the suspicion that the fault lies with the author and that the writing is bad. Alternatively, he may surmise that the author is somewhat confused, does not fully understand the subject, or is not clear about what he is trying to say. An author who knows his subject thoroughly and can write well should be able to make his exposition readily intelligible, even with an abstruse subject.

A presumption of bad writing, however, is only a gut reaction. The reader will have more confidence in his own judgment if he can justify his intuitive feeling through analysis. The remainder of this book should make this task easier.

Writing is a skill, like golf. Some persons are naturally good at it, most are not, but all can improve with practice, especially if guided by proper instruction. In golf we have an objective measure of merit—the score, and the score can also serve as an index of improvement. In writing, as already noted, there is no such objective measure. Nevertheless, good writing does differ from bad, and the more readily an aspiring author can recognize the bad, the more readily can he improve his own writing. He should become sensitized to certain major faults, so that when he encounters them they will engender acute discomfort—as if someone had rubbed sand on a fresh sunburn.

There is a perpetual dialectic in regard to standards—we crave them and yet we rebel against them. The critic—I use the term in its best sense—has the function of analyzing and justifying standards. Each reader of this book, I assume, will develop his own critical faculties and set his own standards. I enunciate my own values, with the hope that readers will agree. The important factor is the remedy—the alternative pathways that can lead to better writing. My approach is the empirical presentation of alternatives and the reasons that underlie them. The reader makes the choice.

Does a given passage read smoothly and easily? If not, how to make it smoother? The precepts that I enunciate I have used in   **19** editing manuscripts and in teaching courses in writing. If the authors (and students) agreed that the emendations resulted in better prose, I was satisfied. In this book I try to communicate the principles that underlie my changes so that the readers can adopt these principles if the changes seem good. A student faces alternatives. His intuitive reactions will usually correlate well with my own values. I try to provide an explicit rational basis for the intuition.

I would mention one important difference between editing a journal and teaching a course in writing. The editor has definite responsibilities, and the carrying out of responsibilities entails the exercise of authority, which may seem arbitrary. However, if a student in my writing course is not convinced that my suggestions bring about an improvement, I will not push the issue. In this book I do not assert any authority. I merely offer alternatives for you, the reader, to accept or reject—after exercising your critical judgment.

3

Five Treacherous Servants

The late Richard Asher, perhaps the finest medical writer of this century, wrote an essay "Six Honest Serving Men for Medical Writers" [1]. The title he took from the familiar verses of Kipling:

> *I keep six honest serving men*
> *(They taught me all I knew):*
> *Their names are What and Why and When*
> *And How and Where and Who.*

I have modified Asher's title to identify certain parts of speech that, when used skillfully, can render prose clear and agreeable and, when used unskillfully, can make the prose clumsy and obscure. I refer to the verb *to be*, and to the preposition, the conjunction, the modifier, and the pronoun. For simplicity I represent them as *is, of, and, very,* and *it.* Properly disciplined they are admirable servants; undisciplined, they can wreck any composition.

In my presentation I interpolate some bits of elementary grammar. Many students, I have learned, are grateful for these reminders.

Is

Verbs express action: In the sentence "He hit the ball," a subject *he* performs the action *hit*, affecting an object, the *ball*. We call *hit* a transitive verb because it takes an object. On the other hand, in the sentence "He ran," the subject also performs an action but does not affect any object. The verb *ran* we call intransitive.

We also distinguish active from passive, thereby referring to a property we call *voice*. This indicates a particular relationship between subject and verb. If the subject is, so to speak, on the delivery end, as in the sentence "He hit the ball," we designate the verb as active or in the active voice. If the subject is on the receiving end, "He was hit by the ball," the verb is passive. The passive voice always includes some form of the verb *to be*, which serves as an auxiliary to the main verb.

To be stands apart from other verbs, for by itself it does not indicate any action. It may serve as an auxiliary, entering into a compound 23

verb form, but any action depends entirely on the main verb. The auxiliary merely shows where the action of the main verb lies. We appreciate this if we recall that in Latin we express the passive voice by a grammatical inflection, without any distinction of main and auxiliary verbs. Where Latin uses a verb ending, English uses an auxiliary.

A further use of *to be* as an auxiliary we find in the so-called progressive or continuing form of a verb. "He was walking" tells us something rather different from the simple "He walked," for it makes explicit the continuity of an action, namely, a progression. In Latin this sense is conveyed through a specific ending, that is, through an inflection of the verb. English, as a poorly inflected language, must use another mode of expressing the same idea, and does so by combining an auxiliary verb with the participle. The latter reveals the activity while the auxiliary replaces the case ending found in the Latin. The progressive forms of a verb—is walking, was walking, had been walking—can thus give precise shades of meaning.

To be also functions as a *copula*, a junction word that joins the subject to another term that the grammarians call the *subject complement*. Thus, in the two sentences "John Doe is tall" and "John Doe is the author," the *is* links John Doe in the one case with an adjective, in the other with a noun. As copula the *is* makes explicit the relationship between subject and complement but does not itself contribute to the relationship, nor does it limit either term, nor, of course, does it express any action.

In the discussion of *to be* I will consider only its function as copula and as auxiliary in the passive voice, for these are the functions whose abuse makes so much prose dull and flabby. I will not take up here the use of the progressive form.

If we want to eliminate a major component of bad writing, namely, the excessive use of *to be*, we must first become sensitive to the deadly quality of the prose in which this excess occurs. Read the following quotation aloud and stress slightly the words in roman. (The quotation refers to the serum calcium levels under certain conditions of hyperparathyroidism.)

The serum calcium level is *constantly elevated. However, there* is *no change in protein binding, and measurement of total calcium* is *as useful as measurement of any fraction. Hypercalciuria* is *usually present, but* is *not useful as a sign, since it* is *abolished if glomerular filtration* is *impaired, as it* is *in many cases of hyperparathyroidism.*

This passage contains eight verbs, all of them embodying an *is*, either as a simple copula or as part of a passive verb.

Take this further example:

This small monograph is *an excellent summary of current concepts of neurophysiology. It* is *profusely illustrated with beautiful pictures and clear diagrams. The text* is *relatively simple and* is *obviously written for the non-expert, for there* are *very few references cited.*

These two sentences contain five verbs, in all of which some form of *to be* appears, twice as copula and three times as auxiliary in passive verbs. I maintain that these examples contain far too many instances of *is* and *are* forms and that if we eliminate most of these we will bring about an improvement: the sentences will be *better*. To indicate some ways to get rid of *to be* forms, I will consider the second example.

Surprisingly often we can simply omit the copula and bring the complement into direct contact with the subject. We may link an adjective directly to the subject and thus eliminate an *is*. Instead of saying "The text is relatively simple," we need merely say, "The relatively simple text" and then attach the subject, *text*, to some other verb. In the present example we say, "The relatively simple text is obviously written. . . ." One *is* has disappeared, along with an *and*.

Another common technique converts a noun complement into a verb: *"is* a summary" becomes *summarizes*: "This small monograph summarizes current concepts. . . ." What, then do we do with the *excellent* that originally modified *summary*? We convert it into an adverb that modifies *summarizes*, so that *excellent summary* becomes *excellently summarizes*. "This small monograph excellently summarizes current concepts. . . ." Another *is* has disappeared, along with the preposition *of*.

A further technique involves some combination and transposition. In the sentence "It is profusely illustrated . . .," the *it* refers

25

to *monograph*, and the sentence tells us about the fine illustrations. If we omit the *is*, we have "It [the monograph], profusely illustrated. . . ." If we now eliminate the *it*, we can attach the *illustrated* directly to the word *monograph* in the previous sentence. This transposition would then give us "Profusely illustrated with beautiful pictures and clear diagrams, this small monograph. . . ." We have eliminated another *is*.

An important technique for changing the passive voice into the active requires us to turn the original subject into the grammatical object. "There are very few references cited" turns into ". . . cites very few references." What, then, is the grammatical subject? What agent does the citing? Obviously, *the text*. But since the word *text* has already occurred in the main clause, we can use the pronoun *it* in the subordinate clause: "for it [the text] cites very few references."

If we put all this together we find that the revision reads

Profusely illustrated with beautiful pictures and clear diagrams, this small monograph summarizes excellently the current concepts of neurophysiology. The text, relatively simple, is obviously written for the nonexpert, for it cites very few references.

We now have three verbs, two of them active and one passive, and no copulas at all. We have eliminated four of the original five instances of *to be*. Is not the revision an improvement over the original?

I have no intrinsic objection to the passive voice nor to the copula. I object only to their excessive use. The copula, instead of adding something of its own, merely fulfills a grammatical need to make a complete sentence. Even though sometimes indispensable, the copula is, so to speak, an "empty" word. Hence, too many copulas dilute the sense. Since empty words automatically exclude vigorous expressive verbs, they may lead to bovine monotony and flabby writing. The passive voice may sometimes provide the exact shade of meaning that the author intends, but overabundant use indicates slipshod habits of writing, combining ignorance, carelessness, and laziness. More rarely, excessive use of the passive voice stems from adherence to standards that in my opinion are utterly wrong, as I will explain later.

What constitutes an excess? I suggest a rule of thumb: In any given segment of writing, no more than one-fourth to one-third of the verbs should be copulas or passives. For those persons willing to make a substantial effort to improve their writing, I suggest the following exercise. Choose eight to ten consecutive sentences from any random text and count the total number of verbs. Record this as the denominator of a fraction. Then count the copulas or passives and record that number as the numerator. If the fraction exceeds ⅓, consider the number of copulas as excessive. If you are examining your own writing, try to eliminate as many of the passives or copulas as you can. (Please note that I do not include the progressive forms of verbs, for these usually contribute to precision and vigor rather than detract therefrom.)

In the following additional examples wherein copulas or passives can be eliminated with advantage, notice one particular warning flag, which I call the *is-and construction*. This we have already encountered. Here is another instance.

> *Man* is *a part of nature* and *shares in the phenomena that apply to all other animals.*

If we delete both the *is* and the *and*, and insert commas in their places, we have an improved sentence. For the sake of euphony we can insert an *as*, so that the sentence reads

> *Man, as part of nature, shares in the phenomena that apply to all other animals.*

Again

> *The second part of the text* is *directed to clinical pathology* and *begins with a general discussion of apparatus.*

Delete both the *is* and the *and* and insert commas, turning the sentence into

The second part of the text, devoted to clinical pathology, begins with a discussion of apparatus.

Or again

The book is divided into three sections and consists of. . . .

becomes

The book, divided into three sections, consists of. . . .

Sometimes, when we start to get rid of an *is*, we find that we can spare other words as well. Thus

The work that is represented in this book is a valuable contribution to physiology and will undoubtedly be widely used as a reference source.

Obviously, we can easily get rid of the first *is*, together with the accompanying *that*, to yield

The work represented in this book is a valuable contribution. . . .

But do we lose anything if we chop a little more? *The work represented in* can readily disappear without loss of meaning and leave us with "This book. . . ." Then we can correct the *is-and* construction, and the original clumsy sentence becomes

This book, a valuable contribution to physiology, will undoubtedly be widely used as a reference source.

Please note that we have not meddled with the passive voice *will be used*.

Another empty construction is the all-too-common usage, "The fact is that," "It is clear that," and similar clauses. These we should eliminate ruthlessly. Sometimes, to the resulting shortened version we can profitably add an adverb to render the sentence more euphonious and provide a better transition. Then a sentence such as

The fact is *that thieves* were *an important class in the total social structure*

becomes

Indeed, thieves were *an important class in the total social structure.*

In regard to copulas, I strongly advocate the technique of changing the noun complement into a verb. In the sentence

This excellent volume is *a compilation of papers presented at a symposium*

we can delete the *is* and turn *compilation* into the verb *compile*. Thus,

This excellent volume compiles the papers presented at a symposium.

But since *compiles* demands a personal subject and *volume* is inanimate, we can change *compile* to *gather* or *bring together*, and say

This excellent volume brings together the papers read at a symposium.

Sometimes the elimination of an *is* demands the insertion of a facilitating word such as *as*, to cover a change in construction.

An unexpected feature of this book is *the inclusion of a discussion on gynecological emergencies.*

This would become

As an unexpected feature, this book includes a discussion of gynecological emergencies.

A sentence in the passive voice tells us that something has been or is being done. To indicate an agent we would need a prepositional phrase. A writer might say, "The book was written by me," a sentence whose meaning is clear but whose clumsiness would be hard to surpass. Yet we have no problem in converting it to the active voice: "I wrote the book."

Many times the passive construction does not tell us the identity of the agent, but allows us to presume it. Take this example, from a book review commenting on a recent volume on statistical method.

The concept of randomization is stressed throughout as a key principle in design. However, the difficulty of achieving a true random sample is not fully discussed.

If we ask, Who does the stressing or the discussing? the answer is obviously, The author of the book. We can improve the review by changing the passive to the active and making the agency explicit. At the same time we can make a few judicious transpositions and insertions.

Throughout, the author stresses the concept of randomization as a key principle in design, even though he does not fully discuss the difficulty of achieving a true random sample.

Often, however, the identity of the agent remains unclear. In medical journals we constantly see statements that a problem was investigated, a patient was transferred, a catheterization was performed, and the like. To answer the question, Who did these things? we need to take account of certain additional factors. If, for example, the author declares, in the opening of his paper, that "the problem of blood flow was investigated," we must gather from the context whether he is referring to his own work—that he is the agent—or to the work of some other investigators. Ordinarily, but not always, the context allows us to make a clear decision. It would be far more helpful, however, if he said, "I [we] investigated . . ." or "Doe and Roe investigated. . . ."

Rather different is the sentence "The patient was transferred to the surgical service." Is the author referring to the physical transport of the patient, presumably performed by an orderly? If so, does it make any difference who actually moved the patient from one place to another? It would be otiose to make an explicit statement, "The orderly transferred the patient to the surgical service." If the author had in mind not the physical transport of the patient but the administrative transfer—the shift in responsibility from the physician to the surgeon—then the agent would presumably be some particular clerk making a suitable notation. But again, there is no point whatever in indicating the agent. The sentence "The patient was transferred to the surgical service" tells us all we need to know, and any effort to change the sentence to the active voice would be counterproductive.

The sentence "He was catheterized" might seem comparable. Indeed, if we are talking about the urinary bladder, it would make little difference who passed the catheter. But if we are talking about cardiac catheterization, we might want to know who actually performed the operation.

When we use the passive voice we must always consider whether the identity of the agent is significant to our story. Consider the following quotation:

> *When the condition of a patient at the time of discharge was suggestive of an undiagnosed wound infection, an effort was made to trace the patient through local nursing homes and family physicians.*

This sentence contains two verbs. The first, *was*, with its complement *suggestive of*, is certainly clear, even if clumsy. The natural mode of expression, however, would be "When the condition of the patient . . . suggested a wound infection. . . ." Torturing this into the passive voice indicates a virtual obsession that I will discuss shortly. In the second verb, "an effort *was made* to trace the patient . . .," the agent has considerable importance. *Who* tried to trace the patient? A trained social worker? A ward clerk? A surgical resident? Perhaps the author himself? Unless we know who made the effort, we have no way of evaluating the results. The passive voice can lead to ambiguity and doubts, as well as to clumsiness and monotony.

How can we account for the popularity of the passive voice? I suggest two major factors. The first relates to defects in our educational system. As various literary journals have complained, nowhere in our educational system are students taught to write. With rare exceptions, teachers and students alike avoid contact with English composition.

The second factor has to do with a mistaken notion about science 31 and its relation to "scientific" communication. The alleged objectivity of science has hypnotized many otherwise capable scientists, who regard anything subjective as tainted, to be avoided as much as defective instruments or contaminated solutions. The logic is simple. The active voice will necessarily require abundant use of the first person—"I did this" or "We did that"; *I* and *we* are subjective, to be avoided as unscientific; the only alternative is the passive voice which, by avoiding the first person, becomes the favored mode of expression.

With this point of view I must disagree in the strongest possible terms. I maintain that objectivity in science is a myth, and that if the devotees of this mythology would apply themselves to clear expression rather than to indefensible dogma, we would have a far greater general benefit.

I am reminded of a seminar in medical writing that I once gave to a group of residents. I pleaded for fewer passive constructions and greater use of the first person. When I had finished, several residents mentioned that the head of the department, who reviewed all manuscripts intended for publication, positively forbade any use of the first person. Everything had to be in the third person, even when a resident was applying for a research grant. Such a blanket rule, as it spreads the illusion of objectivity, also encourages use of the passive. All I can do is to exhort my readers not to follow this example.

Before we leave the verb *to be*, there is one further usage that deserves notice—the so-called absolute construction. This refers to the combination of a noun and a participle that together form a grammatical unit not connected with anything else in the sentence. The construction is most familiar in Latin. Caesar would have said, "The town having been captured, the army crossed the river." "The town having been captured" translates an ablative absolute, *oppido capto* in which the noun in the ablative is combined with a past participle also in the ablative. These two words stand by themselves. This is good Latin, but a literal translation is not good English. An idiomatic translation might read, "After the town was captured, the army crossed the river."

From time to time we find the absolute construction in English. A book reviewer commented on a new edition of a book in a rather technical subspecialty. He declared,

Therapeutics is but little changed from the previous edition, there having been no marked advance in the interim.

"There having been no marked advance" modifies nothing and stands by itself as an absolute. To bring it into relation with the rest of the sentence, we change the absolute to a subordinate clause: "for there have been no marked advances in the interim."

A further example describes a contemporary physician:

Like his colleagues, he was famed for his skillful knowledge of cardiac disorders, his name being associated with heart block.

This we change to ". . . skillful knowledge of cardiac disorders, especially in relation to heart block." There are, of course, many different transformations possible, just as there are different possible translations of a Latin ablative absolute.

The absolute construction is not wrong, merely stilted and clumsy. In my own editing, I always delete it and make some appropriate substitution.

In summary, let us avoid any passionate devotion to the various forms of the verb *to be*.

Of

In a major university a department head addressed a meeting of his graduate students:

The present meeting has been occasioned by an acute sense of horror created in me by the recent perusal of the first drafts of a large number of literary ventures submitted by various members of this department.

This quotation comes not from any off-the-cuff remarks, hastily spoken and literally transcribed, but from a published version that had already been revised. The professor had wanted the students to improve the quality of their writing. If the quotation represents the style of the professor, we may indeed wonder just what sort of writing the students had been submitting.

The sentence in question, with 37 words, rests on a single passive verb, *has been occasioned.* Following the verb we find a chain of nine prepositional phrases, among which are interspersed a few modifiers. If we determine what I call the preposition quotient—the ratio of prepositions to total word count—we get 9/37, or virtually 1:4. Almost every fourth word is a preposition.

Prepositions are connecting links that show a relationship between two other terms, i.e., between the object of the preposition and the word that the prepositional phrase modifies. Use of too many prepositions creates a disproportion between linking words and the verbs, nouns, and modifiers that bring vigor and life to the sentence. This disproportion can render the sentence ill indeed. A ratio of one preposition to every four words is a bad prognostic sign.

The overuse of prepositions is a severe and extremely common fault. Indeed, if I wanted to offer a single rule for improving the quality of writing, I would unhesitatingly say, Reduce the number of prepositions.

How many prepositions are too many? To this question I can give no clear answer, but I will mention some warning indicators. Long sentences that are not easy to understand and that have only one or two verbs, either copulas or passives, should make you suspect that the sentence has too many prepositions. The real test is, Does elimination of prepositions improve the sentence?

To cut down on prepositions, I suggest several techniques. In pointing to certain constructions that lend themselves to change, I do not mean to imply that those constructions should always be changed. If, however, you want to reduce the number of prepositions, here are a few ways to do so.

1. Delete an entire prepositional phrase as meaningless or unnecessary. *"In order to provide* a refuge" means nothing more than

"*To provide* a refuge." By deleting *in order* we have eliminated a preposition.

2. Convert a prepositional phrase into a participle. "*In the attempt* to cross the river" can become "*Attempting* to cross the river." Sometimes such a conversion can kill off two prepositions in one stroke. "*In the fear of reprisal*, he . . ." can become "*Fearing* reprisal, he. . . ."

3. Convert a prepositional phrase to an adverb. "Of the six patients treated *by surgery*, three died," becomes "Of the six patients treated *surgically*, three died." Or convert to an adjective: "It is a question *of importance*" becomes "It is an *important* question."

4. Change the passive voice to the active. "The blood volume *was determined by the technician*" becomes "*The technician determined* the blood volume.

These are simple ways to get rid of prepositions. Other modes may require extensive change in the sentence structure, and cannot be reduced to rules. We may, for example, need to chop one sentence into two, or convert a simple sentence into one that is complex or compound, or provide various transpositions, all depending on the particular example.

Let us now see how we can apply these techniques to some concrete examples.

Let us start by analyzing this sentence and trying to improve it.

There had been major changes in *the presentation related* to *the data accumulated* as *a consequence* of *exhaustive study* of *the results* of *treatment* in *cancers* of *the head and neck, breast, and gynecological tract.*

This example fulfills our criteria for excessive prepositions: a long sentence, a single verb in the passive, and considerable obscurity. There are 35 words and eight prepositions, with a ratio of just over 1:4.

First let us try to clear up the obscurities. To what does *related* refer? Is the *presentation* related to the data, or are the *changes* related to the data? The latter seems to make more sense, i.e., changes (in the presentation) occurred as a result, somehow, of someone studying the accumulated data. Who made the changes? Who studied the data? The sentence, in the passive voice, does not tell us, but we can assume that it was the author of the book.

To begin our revision, we can transfer the passive to the active: "The author made major changes in his presentation." What led him to do so? Presumably he gathered together the results of treatment of the various cases, then made an exhaustive study of these data, and as a result changed his own presentation. If this is correct, how can we express it most clearly? By eliminating prepositions according to the suggestions given above. For example, "*as* a consequence *of* exhaustive study *of*" becomes "*after* exhaustively studying"; "*of* treatment *in* cancers" becomes "*of* treated cancers." The wording "data accumulated *as* a consequence *of*" is a verbose way of saying "accumulated data," but the sentence loses nothing if we omit this entirely. Putting all these alterations together we have

> *The author made changes* in *his presentation* after *exhaustively studying the results* of *treated cancers* of *the head and neck, breast and gynecological tract.*

Here, then, is a simple sentence of 25 words and four prepositions. The revision keeps close to the original but eliminates useless verbiage. If, however, we want to modify the construction, we might say

> *The author changed his presentation after he had exhaustively studied the results* of *treated cancers* of *the head and neck, breast and gynecological tract.*

Now we have a complex sentence rather than a simple one, with 24 words altogether, two verbs in the active voice, and only two prepositions. Is not this an improvement?

Here is a truly grim example:

> From *1969* to *1972, 12 medical centers* throughout *the United States have cooperated* in *a prospective controlled study* of *the effectiveness* of γ-*globulin* in *preventing post-transfusion hepatitis* in *4,210 patients undergoing pulmonary resection.*

This simple sentence, with 34 words, has eight prepositions, three preceding and five following the single verb *have cooperated*. Again we have a preposition index of approximately 1:4.

The sentence is perfectly grammatical. It is, perhaps, not unduly obscure, at least on the second reading, but it lacks all the properties of graceful writing. We can improve it by cutting down the number of prepositions, introducing more verbs or verbals, and rearranging the phrases. There are many possible ways of doing this. I suggest the following:

> *To determine how effective γ-globulin might be* in *preventing post-transfusion hepatitis, 12 medical centers* throughout *the United States have cooperated* in *a prospective controlled study that utilized 4,210 patients undergoing pulmonary resection* from *1969* to *1972.*

We have increased the number of verbs from one to three and made the sentence complex rather than simple. Since this change required additional words to maintain sound grammar, we have actually increased the number of words to 35, but the number of prepositions has diminished to five. The sentence is unequivocally clear; furthermore, it reads perfectly smoothly. We have accomplished this by transforming prepositional phrases into verbs or verbals and rearranging the ideas in a more logical order.

Let us study the following sentence, taken from a context that dealt with the relations between the medical profession and the community.

> *Increasing the professional morale and the skill* in *management* of *illness resulted* in *a pride and prestige* in *the group that lessened the need* for *community regulation* of *moral behavior and technical skill.*

This sentence, whose 33 words include six prepositions, rests on two verbs. The clause "that lessened . . ." has, however, an ambiguous reference. What does the clause modify—did the *group* lessen the need or did the *pride and prestige* lessen the need? Obviously the latter, although the sentence as written allows room for doubt. Except for the ambiguous reference the sentence is entirely grammatical and certainly does not shriek for improvement. Let us see, however, if eliminating prepositions will improve the writing.

Suppose we change "*in* management *of* illness" to "*in* managing illness" and also change "resulted *in*" to another verb that says the same thing but does not need a preposition—e.g., "*induced*"; and then transpose the phrase "*in* the group." The first part of the sentence will then read, "Increasing the professional morale and the skill *in* managing illness induced *in* the group a pride and prestige that. . . ." The ambiguity is gone and we can now work with the last part of the sentence. We have the awkward phrase "*for* community regulation *of.*" The preposition "*for*" has as its object the noun *regulation*, while *community* is a noun-modifier of *regulation*. We eliminate this undesirable construction by changing the phrase "regulation *of*" not to a participle but to the infinitive *to regulate*. Then *community* is the subject of the infinitive, and *moral behavior* its object. The whole sentence, thus edited, then reads

> *Increasing the professional morale and the skill* in *managing illness induced* in *the group a pride and prestige that lessened the need* for *the community to regulate moral behavior and technical skill.*

There are now 32 words and two verbs, but the six prepositions have been reduced to three. The reader should decide whether the revision has yielded a better sentence.

A further comparable example:

> *Dr. Roe and his coauthors are to be commended* for *bringing together* in *one volume the techniques* for *the performance* of *operations* in *all* of *the "anatomic specialties"* of *plastic surgery.*

This sentence, with its single verb, has 31 words and seven prepositions. Improvement is easy. We compress *are to be commended* into the adverb *commendably*, but this leaves us without any verb at all. We get the verb back by changing the phrase *bringing together* into the verb form, *have brought together*. We reduce "techniques *for* the performance *of* operations" to the simple *operative techniques*. In the phrase "*in* all *of* the 'anatomic specialties',*" the *of* is totally redundant and should be omitted. The entire sentence then reads

> *Commendably, Dr. Roe and his coauthors have brought together in one volume the operative techniques* in *all the "anatomic specialties" of plastic surgery.*

There are now 23 words, one active verb, and only three prepositions.

Here is a sentence that I would characterize as truly foggy:

Since biotransformation may result in *destruction* of *the biologically active form* of *a chemical or formation* of *a more toxic product, it is apparent that the presence or absence* of *such mechanisms* within *members* of *a species will effectively alter the concentration* of *the parent chemical* in *the biological specimen.*

The 40 words in this sentence include nine prepositions and the three verbs *may result, is,* and *will alter.*

What is the author trying to tell us? I believe the message is that biotransformation can alter the concentration of chemicals in biological specimens; and the sentence also tells us the reasons why this is so. We can readily improve this. We can change the phrase *"in* destruction *of"* to *destroy.* Thus, instead of "Biotransformation may result *in* the destruction *of* . . ." we will have "Biotransformation may destroy. . . ." We can similarly change "formation *of* a more toxic product" into "form a more toxic product." However, since we already have the word *form* as a noun, we should not use the same word as a verb but should seek some synonym, such as *yield*—"[Biotransformation may] yield a more toxic product." Then, to improve the sentence, we can delete without loss the words "it is apparent that the presence or absence of such mechanisms within members of the species," which merely cloud the sense without contributing anything. Since this deletion also eliminates both the grammatical subject and the main verb, we can make "biotransformation" the grammatical subject of the sentence and use *"will alter"* as the main verb. We then have

Biotransformation, which may destroy the biologically active form of a chemical or yield a more toxic product, will effectively alter the concentration of *the parent chemical* in *the biological specimen.*

We now have only 30 words, three verbs (*may destroy, [may] yield,* and *will alter*), and only three prepositions.

I have a rule of thumb that no simple sentence should have more than four prepositions, and not more than three prepositional   39 phrases consecutively. If there are more than four altogether, or more than three in sequence, I eliminate at least one. The following example, to illustrate this point, has the single verb *is* and five prepositions, four of them in a chain.

The purpose of *Dr. Roe's book is the scrutiny* of *the events attributed* to *the activities* of *the agents* of *various foreign governments.*

By this time the words "is the scrutiny of" should irritate the reader and call loudly for simple relief. We eliminate the *is* and change *scrutiny* to a verb. *Scrutinize* sounds rather precious and might better be replaced by *examines*, or even *examines carefully* (to get the full flavor of *scrutiny*). What, then, do we do with *the purpose of*? Simply omit it. The sentence would then read

Dr. Roe's book examines carefully the events attributed to *the activities* of *the agents* of *various foreign governments.*

If the reader feels strongly about retaining some sense of "purpose," he can say, "In this book, Dr. Roe wants to examine. . . ." I personally would not say this but I would not argue with those who might prefer it.

Ordinarily, when we diagnose sentences as having too many prepositions, we will want to institute appropriate changes and yet preserve the original sense. But from time to time we will come across sentences like this:

For *this reason an awareness had developed* in *recent years that research design directed* toward *an evaluation* of *the relative contributions made* by *different types* of *basic processes* to *the functional losses characteristic* of *age is the "sine qua non"* of *the rapid development of biological understanding* of *the aging phenomenon.*

This sentence has 50 words, two verbs, and eleven prepositions. Obviously too many. But how can we remedy this? Under ordinary circumstances we would first try to analyze the meaning. But here I find myself utterly baffled, for I simply cannot understand what the author is trying to say. Every once in a while, as I go back repeatedly to this sentence, I think I may have grasped at least part of the meaning, but I cannot be sure. In a case like this I would simply discard the whole sentence. And if there were many sentences like it, I would discard the entire manuscript.

No discussion of prepositions should omit the topic of faulty placement, that is, putting a phrase in a position where its reference is not clear and the meaning is confused—or sometimes absurd. The classic example, which I learned in grammar school, is: "Uncle John went out to feed the cow with an umbrella." This obviously calls for a transposition: "Uncle John went out with an umbrella to feed the cow." As a general rule we adopt the principle of placing together those words, phrases, and clauses that belong together, that is, are related to each other in meaning. *With an umbrella* belongs with *went*, not with *feed*.

A somewhat more complicated example occurred in a description of surgery before the advent of anesthesia. The scene is an operating theater early in the last century. Medical students, who had never before seen an operation, were about to witness an amputation. The author went on to describe

> *the nervous pallor of the medical tyros, who are about to see a man's leg or arm whipped off for the first time. . . .*

As the sentence stands, the reader might want to inquire what would happen when the arm (or leg) gets whipped for the third or fourth time. At present, the phrase *for the first time* modifies *whipped*, whereas the author intended it to refer to *see*. All we need do is transpose the phrase closer to the word it modifies:

> *the nervous pallor of the medical tyros, who are about to see for the first time a man's arm or leg whipped off. . . .*

Even better, I believe, would be

> *the nervous pallor of the medical tyros, who for the first time are about to see a man's arm or leg whipped off. . . .*

41

Here is a more difficult example:

The function of the physician who is not a psychiatrist in the treatment of patients with mental or emotional problems has become the subject of a number of studies.

The meaning is quite obscure. What does "psychiatrist in the treatment of patients with mental or emotional problems" mean? So far as I can tell, nothing. That chain of phrases, in strict grammar, modifies *psychiatrist*, but does not do so in any meaningful sense. When we study the sentence, we realize the phrases probably refer to *function*. Transpose the whole chain of phrases and we get a meaningful sentence:

In the treatment of patients with mental or emotional problems, the function of the physician who is not a psychiatrist has become the subject of a number of studies.

This does make sense.

The recommendation, Place together terms that belong together, applies not only to prepositional phrases but also to other constructions. In this connection I would offer one further example, involving a gerundive. The example comes from a newspaper report:

The bride was given away by her father wearing a gown of heavy white lace with a short train. . . .

Need any more be said?

And

Conjunctions, as the name indicates, join together two or more terms, whether these be words, phrases, or whole clauses. Conjunctions are of two sorts. Especially important for this chapter is the class called coordinating—*and, but, or*. The second type we call subordinating—*if, as, when, because*. Let us first consider the coordinating conjunctions.

These join together two terms of equal standing, like two horses yoked together in parallel. The compounding or joining can take place at various grammatical levels. We can have a compound sentence, in which two or more independent clauses are joined together by a coordinating conjunction: "John went home *and* Mary followed." We can have a simple sentence containing a compound subject and a single verb: "Jack *and* Jill went up the hill"; or a single subject with a compound verb: "Jack fell down *and* broke his crown." A verb may have a compound object: "He distributed praise *and* blame"; and so too may a preposition: "He spoke kindly to his nephews *and* nieces." We can have a conjunction of prepositional phrases—two phrases in parallel construction, joined by *and*: "He walked across the hall *and* into the bedroom." "Across the hall" is a prepositional phrase and so is "into the bedroom." They both modify the verb "walked," while the *and* shows their togetherness. We can place adjectives, adverbs, or gerundives into a similar relationship: "The night was calm *and* peaceful." "He spoke clearly *and* distinctly." "Getting *and* spending we lay waste our powers." In all these examples the terms joined by *and* are grammatically coordinate.

Coordinating conjunctions, then, join terms that are equals— neither limits the other in a grammatical sense. The terms thus joined may be complete sentences or parts of sentences—nouns, verbs, adjectives, adverbs, prepositional phrases, gerunds, and gerundives—indeed, any grammatical element that permits a junction.

Subordinating conjunctions are rather different. Instead of equality they show inequality, a relationship of dependence and of limitation. Let us study a simple example:

I will go if *it does not rain.*

This consists of a main clause, "I will go," independent and capable of standing alone but limited by a second clause, "*if* it does not rain." Since this clause cannot stand by itself, we call it dependent. It must attach to something else, and in so doing it alters the meaning of the clause to which it is attached. "I will go" means one thing; "I will go *if* it does not rain" means something quite different.

The meaning of the sentence will vary according to the conjunction:

I will go if *it rains.*
I will go after *it rains.*
I will go because *it rains.*

All have different meanings, but only the conjunction has changed.

We identify sentences according to kinds of conjunctions. A sentence is compound when it has two or more independent clauses joined by a coordinating conjunction. A complex sentence has a main clause and at least one subordinate clause (but it may have more than one). We call a sentence compound-complex when it has at least two independent clauses, conjoined, and at least one dependent clause.

In this chapter, concerned chiefly with parallelism, I use *and* as the prototype for all coordinating conjunctions. Whatever I say about *and* applies equally well to the others. I will not discuss the problems of complex sentences.

And should connect terms that are grammatically similar, that is, that exhibit a parallel construction. When parallelism exists, the conjunction of terms is grammatically sound (even though it may be clumsy). When parallelism does not obtain, the conjoint terms will lead to varying degrees of awkwardness and confusion, even to massive bafflement.

Young children often tend to string their thoughts together with a succession of *and*s, thus creating a single interminable sentence whose separate components have lost touch with each other. We can imagine a child speaking thus:

> *My mother told me to go to the store on my way to school* and *so I dressed quickly* and *ate my breakfast* and *after breakfast I started to school* and *after walking two blocks I forgot what I was supposed to get* and *so I had to go home* and *I asked her again* and *she was mad at me.*

This sentence could go on indefinitely and yet remain grammatically correct (or, let us say, not incorrect). Every word, phrase, and clause will parse without confusion, yet it is hardly graceful. We should improve it markedly if we lysed a few of the *and*s and thus dissolved some of the bonds that tie the clauses together. Yet this behemoth sentence, however ungainly, does have the virtue of parallelism.

I will give several examples of faulty usage, some glaringly obvious, others more subtle. The reader should acquire a sensitivity, so that wrong constructions will act as an irritant and provoke acute discomfort. At first he may not be able to pinpoint the fault or know how to correct it. But if he realizes that something is wrong and tries to analyze the grammar to identify the precise error, then correction should be relatively easy.

My first example comes from an advertisement, printed in three lines, with a broad space between the second and third. The ellipsis dots are part of the original advertisement.

For your many patients
who should avoid aspirin . . .

and *for when they catch colds.*

What does the *and* connect? Not the two naked prepositions *for,* but rather the two prepositional phrases. In judging parallelism we must consider the total *terms* that are united. In this instance each term is a phrase introduced by the preposition *for,* but the two individual phrases do not have a parallel structure. "For your many patients" is quite clear. The preposition governs an object, the noun *patients.* The second *for,* however, is followed not by a noun object but by a clause, "when they catch colds." "For your many patients" and "for when they catch colds" are discordant, and to connect them by *and* is grating indeed.

What is the ad really trying to say? Presumably, that the drug is useful for two classes of patients—those who should avoid aspirin and those who have caught cold. These two thoughts are not coordinate. They demand separate sentences. In this instance, I merely point out the fault. I will not try to correct it.

Here is a confusing sentence whose faults are due to lack of parallelism:

The first part of the book describes operations on the adrenal gland, for tumors that produce endocrinologic changes, and *also to remove the gland for patients with metastases from breast carcinoma.*

In this clumsy sentence the verb *describes* has as its object the noun *operations,* and the writer is distinguishing two different types of 45 surgery: "operations . . . for tumors" and "[operations] to remove the gland." In this original example the *and* tries to join a prepositional phrase, *for tumors*, and an infinitive, *to remove*. These are not parallel. To make the sentence grammatical we should have either two prepositional phrases or two infinitives. It seems easier to have two infinitives. Thus,

> *The first part of the book describes operations on the adrenal gland, to remove tumors that produce endocrinologic changes,* and *to take out the gland in patients with metastases from breast carcinoma.*

The sentence, although still nothing to be proud of, is at least grammatical.

Another example:

> *He had fallen in love with* and *married a girl who had worked in a department store.*

At first glance we might think that the *and* connected the two verbs, but this is not correct. If we break the sentence down into its components, we find "He had fallen in love with a girl who . . ." and "[he] married a girl who. . . ." The verb *had fallen*, intransitive, does not take an object, and to complete the sense a prepositional phrase is added; however, *married* is a transitive verb with the object *girl*. In this example the single noun, *girl*, is performing two separate grammatical functions—the object of the preposition *with* and the object of the verb *married*. We can rectify the fault by having separate objects for the preposition and for the verb:

> *He had fallen in love with a girl who worked in a department store* and *had married her.*

"He had fallen in love . . ." and "[he] had married her" are now grammatically parallel and may appropriately be connected by *and*.

Here is a comparable example:

> *Doctors Doe and Roe have written an excellent account of the chemistry of collagen which probably far surpasses* and *is unlike any other treatise on the subject.*

If we analyze what this means, we find three separate concepts: The two doctors have written an excellent account; this account surpasses any other on the subject; and this account is unlike any other on the subject. The *and*, in the sentence as given, connects the last two thoughts. But grammatically we run into difficulties, for the two clauses have in common the word *treatise*, and this word has a different construction in each clause. *Surpass* is a transitive verb and properly takes an object, namely, *treatise*. However, *is* is a copula that takes not an object but a complement—in this case the phrase "unlike any other treatise." Here, *treatise* is the object of the preposition *unlike*. In the sentence as given the single word *treatise* thus serves as object of both a verb and a preposition, a lack of parallelism made possible by the misuse of *and*.

We can supply a simple remedy by merely deleting the term "*and* is unlike":

> *Doctors Doe and Roe have written an excellent account of the chemistry of collagen, which probably far surpasses any other treatise on the subject.*

But this revision has a possible ambiguity. What is the reference of the pronoun *which*—*collagen* or *chemistry* or *account*? We can eliminate the ambiguity by recasting the sentence into a still better form:

> *In their excellent account of the chemical aspects of collagen, Doctors Doe and Roe have written a treatise which probably far surpasses any other on the subject.*

When the misuse of *and* gets us into trouble, we can often rescue the sentence by recasting it to eliminate the *and*.

The next example, using the conjunction *and* three times, is more difficult. There is a further complication from the use of a noun-modifier. The quotation comes from a book review, in which the reviewer discusses the purpose of the authors:

> *The text is written for the physician involved with birth control services and to help train clinicians and staff in IUD insertion, removal, and patient management.* [IUD stands for "intrauterine device."]

Let us start by analyzing the fault attending the first *and*. It connects two thoughts: "the text is written for the physician involved with birth control services" and "[the text is written] to help train clinicians. . . ." The *and* joins two nonparallel terms, the prepositional phrase "for the physician" and the infinitive "to help." The remedy can take one of three forms: make both terms prepositional phrases, make them both infinitives, or get rid of one of them and thus eliminate the need for parallelism. In the present instance the last course seems best. We can say,

> *The text, written for the physician involved with birth control services, will help to train clinicians. . . .*

In the original sentence, the second *and* is correctly used. It joins *clinicians* and *staff*, terms entirely correlative.

The third *and* involves severe confusion. There are three terms, *insertion, removal,* and *patient management*. The original sentence would imply that these three are parallel, whereas actually they are not. We realize this if we consider the modifier *IUD*. If parallelism existed, IUD would modify all three nouns equally. But clearly, it does nothing of the sort. *IUD insertion* and *IUD removal* make good sense. But *IUD patient management* does not. *Insertion* and *removal* revolve around one thought, while *management* refers to a different thought. The reviewer wants to say that the book will help to train clinicians in IUD insertion; that it will help train clinicians in IUD removal; and that it will help train clinicians in patient management. But that is not what the sentence actually says.

We can eliminate the confusion if we eliminate the noun-modifier in favor of a prepositional phrase. We solve the problem if we say that the book will "train clinicians in insertion and removal of intrauterine devices, and in patient management." Inserting the extra *in* shows that the phrases *in insertion* and *in management* are parallel, both modifying the verb *train*. Hence, the phrases are properly connected by the conjunction *and*. The corrected sentence would then read

> *The text, written for physicians involved with birth control services, will help [to] train clinicians and staff in the insertion and removal of the IUD, and in patient management.*

This is far from elegant, but it is at least grammatical.

The misuse of *and* can produce monstrosities that almost defy repair. Here is a sentence whose context has to do with the care of emotionally disturbed patients.

Physicians reported that more than half of the emotionally disturbed patients were as easy to work with, required no more time to treat, and *showed as good or better improvement than other patients.*

This sentence offers a series of comparisons between patients that are emotionally disturbed and those that are not. All the comparisons lead to the clause "than other patients [with a verb understood]." But clearly, some of the comparisons require the conjunction *as*, while others require *than*. Some require the verb *were*, understood; others the verb *did*, understood. Thus, we should say

as easy to work with as *other patients [were]*
required no more time than *other patients [did]*
showed as good improvement [ugh!] as *other patients [did]*
showed better improvement [ugh!] than *other patients [did]*

The original sentence, by compressing a lot of quite different constructions and running them together in reckless fashion, has produced a grammatical disaster quite hopeless to repair. The sentence must be completely recast with entirely different constructions. I would suggest the following:

Physicians reported that more than half the emotionally disturbed patients responded just as quickly to treatment as did the others, and *improved at least as rapidly.*

This is entirely grammatical. The *and* connects the strictly coordinate verbs *responded* and *improved*. However, the reader may want to recast in a different fashion.

Many times we have sentences that are grammatical and really do exhibit parallelism but are awkward or even utterly absurd. Here is an example of the latter category

Forty percent of all women seeking abortions were married to and impregnated by their husbands, according to Dr. Doe.

The noun *husbands* serves as the object of two different prepositions, *to* and *by*. The conjunction *and*, however, does not connect the prepositions but rather the two verbs, each with its modifying prepositional phrase. Moreover, the sentence is absurdly redundant, for to be married means to have a husband, and it is not possible for a woman to be married to anyone other than her husband. We rescue the sentence by deleting the words *married to and*, with this result:

Forty percent of all women seeking abortions were impregnated by their husbands, according to Dr. Doe.

Here is a further example of severe awkwardness resulting from faulty usage. The context has to do with certain research reports.

Many of the contributions deal with gastrointestinal epithelium because of the ease with which it can be studied in the laboratory and also its inherently high replication rate.

What does the *and* connect? The sentence has three separate thoughts: the contributions deal with gastrointestinal epithelium, this epithelium can be easily studied in the laboratory, and this epithelium has a high replication rate. The thoughts are causally related: because there is a high replication rate, the epithelium is easily studied, and because it it easily studied, there are many contributions on the subject. In the original sentence the *and* connects two nouns, *ease* and *rate*, both objects of the same preposition *of*—once expressed and once understood: "because *of* the ease" and "[because *of* the] inherently high replication rate." But these two objects are widely separated.

The sentence would be improved if, instead of using two widely separated nouns as the objects of a single preposition we substituted verbs. Furthermore, we could place the thoughts in a more logical order to indicate better the causal relationship. I suggest this revision:

Many of the contributions deal with gastrointestinal epithelium because it has an inherently high replication rate and *can be readily studied in the laboratory.*

The subordinate clause introduced by *because* has a single subject, *it*, and a compound verb, *has* and *can be studied*, joined by *and*. We have eliminated useless words and made the sentence more compact and more logical.

Here is another, somewhat comparable, example. The context concerns the curriculum in a medical school.

The number of hours allocated to and *the form of these courses varied widely.*

The *and* would seem to connect *allocated to* and *form of*, but grammatically this is not the case. The *and* actually connects the two nouns, *number* and *form*. The number of hours varied widely and the form of the courses varied widely. Thus the one verb has a compound subject *number and form*. But then the author has introduced some modifiers. *Number* is modified by the phrase *of hours*, while *hours* is modified by the participle *allocated*; *allocated* is modified by the phrase *to these courses*. The other half of the subject, *form*, is modified by the phrase *of these courses*. The one noun *courses* serves as object in two different prepositional phrases that are not parallel—one phrase modifies a participle and thus is adverbial in nature, the other modifies a noun and is adjectival.

To straighten out the difficulty I suggest two possibilities. First, retain the compound subject and the single verb, but give each prepositional phrase a separate object. A rendition grammatically correct but dreadfully clumsy would be

The number of hours allocated to these courses and the form of the courses varied widely.

Some minor changes would slightly improve the flow but retain the same format:

The form of these courses and *the number of hours allocated to them varied widely.*

However, the sentence would be vastly improved if, instead of a simple sentence with a compound subject, it was compound:

The number of hours allocated to the courses varied widely and *so did their form [vary widely, understood].*

The *and* now connects two independent and coordinate clauses.

In contrast, I offer this example:

The information was passed along to and *shared by the hospital staff.*

This sentence has a single subject, *information*, and a compound verb, each part of which has its modifiers. The information, we learn, was passed along to the hospital staff, and the same information was shared by the hospital staff. We note that the two prepositional phrases, modifying separate verbs, nevertheless have a single object, *staff*. However, each phrase is adverbial in character, and the two are strictly parallel.

In this sentence, what does the *and* connect? Not the verb forms, not the prepositional phrases, but the two halves of the compound predicate. These halves, strictly coordinate, exhibit parallelism; the *and* is appropriately used. However, although the sentence is grammatical, I consider it clumsy and undesirable. I would revise the structure so that two prepositions do not govern the same object. I believe the sentence is much more pleasing if we change the construction and say,

The hospital staff shared the information which had been passed along to them.

This gives a complex sentence with a subordinate clause, instead of a simple sentence with a dual predicate. The *and* has disappeared. However, the original format was not wrong, and those who like it better than my version are welcome to their preference. Some people like gooseberries, others don't.

Regardless of preference in any individual case, I would recommend a general rule in regard to *and*: every time you use it, ask yourself the question, What does it connect? If you cannot give an instant and clear answer, make some changes in your constructions.

Very

I use the adverb *very* to exemplify the general class of modifiers. A modifier limits the meaning of some other term. The most common modifiers are adjectives, which modify nouns; and adverbs, which modify verbs, adjectives, and other adverbs. Take a random noun—say, *boy*. It indicates any male child without distinction; add an adjective, like *tall*, and you immediately create a limit. Tom, Dick, and Harry are all boys, but only Tom is tall.

Adjectives have a descriptive function. We can build up a picture by piling one adjective on top of another—one boy is tall, strong, handsome, and generous; another is tall, spindly, feeble, and spiteful. These various adjectives do not react on each other but all modify the noun *boy*. Adverbs, however, among their other functions, can modify adjectives and limit their meaning. *Very, slightly, exceedingly,* and *surprisingly* will each change the meaning of the adjective in a different way. "The boy is very tall" conveys a picture quite different from "The boy is surprisingly tall."

Adverbs modify verbs: "He walked quickly" or "He walked slowly." Furthermore, some adverbs can also modify other adverbs. "He walked surprisingly quickly" is not at all the same as "He walked quickly."

The class of modifiers also includes prepositional phrases which may limit the meaning of either nouns or verbs. "The girl with the blonde hair" contains an adjectival phrase. This can itself contain an adverbial modifier: "The girl with the artificially blonde hair." There are also adverbial phrases: "He cried out in a loud voice." Sometimes such a phrase may be replaced by a simpler modifier that conveys the same limitation: "He cried out loudly." The adverb *loudly* is more or less the equivalent of the adverbial phrase "in a loud voice."

As we saw in the discussion of conjunctions, subordinate clauses also exert a modifying or limiting function. In the complex sentence

"The boy is tall when he stands up straight," the subordinate clause "when he stands up straight" limits the meaning of the main clause "the boy is tall." In this section I will discuss a few of the problems that involve modifiers but will omit all considerations of prepositional phrases and limiting clauses.

Modifiers, then, qualify and limit the terms to which they apply. Badly used, they can render writing flabby. It is easy to make a general rule, "Avoid excessive use of adjectives (or adverbs)," but such a rule has singularly little value. How much is too much? How many adjectives should a sentence have? Although we cannot answer such questions, we can identify certain contexts that do have too many modifiers (and sometimes even one modifier is too many).

Let us start with the word *very*, which provides the heading for this section. This word derives, ultimately, from the Latin *verus*, true; and some sense of true or truth attaches to the various meanings. *Very* is both an adjective—"the *very* end," "the *very* idea"— and an adverb that modifies an adjective (to indicate a higher stage or degree). We may call it an intensifier. "A beautiful picture" has one meaning; "a *very* beautiful picture" indicates a higher grade of beauty.

Beautiful is an adjective so hackneyed that it has lost all force and really indicates little more than mild to moderate approval. When an adjective becomes merely vapid instead of expressive, we may try to restore some of its lost vigor by adding *very*. If the mind no longer reacts to a verbal stimulus, the adverb *very* tries to increase that stimulus. However, since the mind will soon adapt to that added stimulus, then further intensification will be necessary if we want to attract attention. In movie advertisements each new film may carry an epithet more extravagant than its predecessor. *Very* is far too pale. If one picture is *stupendous*, the next will be *colossal*, and the next *supercolossal*. But such extravagance carries the seeds of its own destruction. A few years ago this story was making the rounds: A Hollywood producer described to a friend the latest picture from his studio. "It is marvelous, tremendous, breathtaking, colossal—why, it might even be *good*."

The adverb *very* has become emasculated through overuse. I

would make this suggestion: Every time you have the impulse to write *very*, repress the impulse and see whether the omission would entail any real loss of meaning. At first you will almost certainly say, Yes, something *is* lost. But I predict that if you persevere, you will gradually agree that any loss is negligible. *Very* is one of the words that contributes to flabby writing.

Adjectives, when overused and tired, contribute but little and may even detract. This is especially noticeable when virtually every noun is modified by an overworked adjective. Take, for example, the following advertisement:

> *The world*-famous *Jumbo cruise offers an* unusual *opportunity to enjoy a* rare *close-up view of the* magnificent *estates and* lavish tropical *gardens which line the* secluded private *islands and* beautiful *waterways of Doolittle Beach. Here you will have the* rare *opportunity of visiting the* magnificent formal *gardens, studded with* sparkling *fountains,* crystal *pools,* beautiful *statuary and* sculptured *colonnades.*

Almost every noun has its adjective, and the net result is tiresome and dull.

Overloading with adjectives was at one time perfectly acceptable. Indeed, adjectives, if forceful, may provide considerable impact. Here is one example, written in 1849, that described the social environment in the slums where a cholera epidemic flourished [4]:

> *It is in a nation's dens of poverty, where* unrequiting *toil pines for its* daily *food, where nakedness shivers in the* wintry *air, where the* miserable *victims of* unjust *conditions of society hive together in* damp *cellars and* unhealthy *garrets, where the* blessed *air of Heaven is tainted by* unventilated *streets and* dark *and* obscure *alleys, where* pure *water is a luxury. . . .*

To the modern ear these words sound stilted. The quotation is not without force, but the effectiveness derives as much from a few striking verbs—*pine, shiver,* and *hive*—as from the overabundant adjectives.

Here is a further example from a little-read Victorian novel, *Coningsby,* by Disraeli. He was writing about youthful leaders in politics who might in time lose their idealism [3].

> *[Will] their enthusiasm evaporate before* hollow-hearted *ridicule, their* generous *impulses yield with a* vulgar *catastrophe to the* tawdry *temptation of a* low *ambition? Will their* skilled *intelligence subside into being the* adroit *tool of a* corrupt *party?*

And the rhetoric goes on and on. Many of the adjectives, individually, are forceful and by no means hackneyed, but the combination of adjective and noun, adjective and noun, in monotonous sequence, is both pompous and wearing.

On the other hand a transposition of modifiers may restore freshness and vigor even when the adjectives themselves are trite. The same author describes a reunion of old schoolmates [3]:

> *And yet there is perhaps no occasion when the heart is more* open, *the brain more* quick, *the memory more* rich *and* happy, *or the tongue more* prompt *and* eloquent, *than when two school-friends meet.*

We can appreciate the effectiveness if we reverse the position of adjectives and nouns into the more usual order:

> *no occasion when there is a more open heart, a more quick brain, a more rich and happy memory, or a more prompt and eloquent tongue.*

This is dull, whereas the original version had some sparkle and rhythm. The placing of the adjective after the noun instead of before it is a most helpful technique.

Overloading with adjectives is a fault common in medical writing. Here is a simple example from a book review:

> *The* excellent *chapters on* legal *principles contain warnings about* mandatory *testing, which appear in* muted *form in the* terse *recommendation section but could bear* added *emphasis.*

This passage contains eight nouns, of which one, *recommendation*, serves as a modifier, the so-called noun-modifier. Of the remaining seven nouns, six are preceded by adjectives. Only *warnings* remains unmodified. Yet despite the overload of adjectives, the sentence has an important merit: Of the 26 words, three are verbs. These tend to mask the numerous adjectives, so that they do not seem overly intrusive. Compare it with the following sentence, concerned with the lungs in emphysema.

> *Specimens containing* progressive *stages of* minimal *to moderately* advanced, subclinical, localized *but pathologically* typical, obstructive pulmonary *emphysema were utilized for the study of* early *and* developing bronchiolar *and* respiratory *tissue lesions.*

Here are 31 words, resting on a single verb. Apart from the subject, *specimens*, there are four nouns functioning as such (*tissue* is a noun-modifier and therefore functions as an adjective) and twelve adjectives, including the gerundive, or verbal adjective. In addition, two adverbs modify the adjectives, so that altogether there are fourteen modifiers (fifteen, if we include "tissue") bearing on four nouns, which in turn rely on a single verb. Obviously the sentence has excessive modifiers.

Here is another instructive example:

> *This* fluent, *highly* readable first English *translation of John Doe's* famous *monograph will be of* special *interest to those interested in the* philosophical *aspects of sociology.*

Preceding the noun *translation*, we have four adjectives and an adverb, five modifiers in all, while of the additional four nouns three have single modifiers.

This reminds us of the Germanic constructions which often encourage long strings of modifiers preceding a noun. Here, for example, is a literal translation from the German:

> ... *otherwise the on-the-spit-grilled, in-olive-oil-immersed, with pepper-and-salt-thyme-and-mustard-seasoned, and with brown-butter-poured-over, extra-fine lamb kidneys.* ...

But good or at least acceptable German usage does not make good or acceptable English.

Not all authors are sensitive to the differences between German and English. What, for example, can we make of this excerpt:

The average resting and after arginine hydrochloride infusion plasma growth hormone concentration of relatively coronary-prone subjects. . . .

In fairness to the authors we must realize that this comes from a synopsis-abstract, in which the number of words was strictly limited. The authors tried to crowd as much information as possible into a limited number of words. But merely compacting a series of words, without regard to style, produces not a communication of thought but an indigestible lump.

Some modifiers are forceful. We think of them as possessing vigor; they produce a powerful effect; they call to mind unusual imagery, create associations that lend richness and that fit into a pattern. However, when a modifier becomes trite it loses its force, produces no effect, and even detracts. Masters of prose style avoid piling up commonplace terms which communicate nothing except a sense of intellectual poverty. Let me give an example of superb use of modifiers. Carlyle is describing his thoughts of death [2]:

And yet, strangely enough, I lived in a continual, indefinite, pining *fear;* tremulous, pusillanimous, apprehensive *of I knew not what: it seemed as if* all *things in the Heavens above and the Earth beneath would hurt me; as if the Heavens and the Earth were but* boundless *jaws of a* devouring *monster, wherein I,* palpitating, *waited to be devoured.*

The modifiers are of two sorts, adjectives and gerundives (or verbal adjectives). The latter, like *pining, devouring,* and *palpitating*, convey the force of their verbs, thus adding an especial vigor. Many of the adjectives precede the words they modify, many others follow (where the modified word is the personal pronoun *I*). This shift in position prevents the monotony we will see in the next example. Notice that the modifiers have more color and force than do the verbs. The verbs *lived, knew seemed, hurt, were,* and *waited* are all precise enough, but they do not convey any special imagery.

58

In this sentence the verbs provide the framework that carries the modifiers.

Let us go from a master stylist to an expert in bureaucratic gobbledygook:

> *It is a* virtual *certainty that the* spatial *pattern of a city in a* free-enterprise *society is the* collective *result of a* large *number of* separate business *and* household location *decisions and* transporation *choices.*

In this quotation I have placed in roman the modifiers to which I want to call attention. These include adjectives and also nouns, the so-called noun-modifiers. There are no gerundives to lend the force that inheres in verbs. Furthermore, all the modifiers precede nouns. Indeed, if we disregard the noun-modifiers (which are functioning as adjectives) then *city* is the only noun that does not have a modifier preceding it. The prose is lifeless and depressing. Of course, the abuse of modifiers is only one of the faults here. Others include the dispensable clause "It is a virtual certainty that" and, when that is eliminated, the excessive weight that rests on the single verb *is.*

We can digress for a moment and consider the vexing problem of noun-modifiers. In the example quoted the author talks about *free-enterprise society.* This is a not uncommon usage and perhaps is even preferable to the rather clumsy *society devoted to free enterprise.* But we cannot be so tolerant of *business location decisions, household location decisions,* and *transportation choices.* A single noun-modifier in a sentence might be tolerable, but a series of them is disastrous, especially when two nouns join to modify a third.

Some purists insist that nouns should never serve as modifiers, but this ignores established custom. We constantly speak of heart disease, science fiction, and virus titers. To be sure, we might say disease of the heart, or titers of the virus, or (with marked distortion) scientific fiction, but these would sound stilted. We should also note that a person who studies law is a law student and not a legal student, even though a person who studies medicine is a medical student and not a medicine student.

Noun-modifiers do have a definite, limited place, established by usage, but the limitations are too often transgressed. Take for example,

Those who address the issue of drug abuse treatment evaluation. . . .

Here we have four nouns in a row, three of them serving to modify the fourth. In this monstrosity there is a hidden difficulty, namely, a tautology. If we leave out the noun-modifiers, we see that the quotation concerns "those who address the issue of evaluation." But this is only a long-winded way of saying "those who evaluate," or even "those who want to evaluate." If we get rid of the useless words, we could readily say,

Those who evaluate the treatment of drug abuse. . . .

This, I believe, is what the author really wanted to say. But, as I emphasize in another chapter, an author usually does not know what he really wants to say until after he has said it clumsily two or three times; and when he recognizes that his exposition is thoroughly bad, then he may be able to cleave to the essence and express his message more gracefully.

I suggest that nouns acting as modifiers should be followed by a hyphen, to indicate a unity-in-duality. We are so accustomed to *heart disease* that a hyphen seems unnecessary. But the hyphen removes the first noun from the category of modifier and makes a single compound noun. *Heart-disease* is one noun. In *heart disease* we have two nouns, the first of which is a modifier. However, as a realist I do not really expect the suggestion of hyphenation to take root and prosper.

We must be careful about chains of modifiers, even when they are legitimate adjectives. Careless writers often will run together a series of adjectives, some appropriate and others quite inappropriate. Here is a sentence that plays fast and loose with modifiers. A critic was describing a dance recital.

These limber smart understated *yet* joyous *dancers rival what I have seen any* seasoned major arts *institution achieve on a London stage.*

Dancers may indeed be limber, smart, and joyous, but what does *understated dancer* mean? Presumably the writer had in mind that the dancers did not exaggerate their activities but exhibited restraint and other comparable qualities. But then only the dancing was understated, not the dancers. The rest of the quoted sentence is indeed clumsy, and we would have great difficulty if we wanted to figure out exactly what the dancers are rivaling, but this is a problem distinct from the misuse of modifiers.

Some writers, not content to say a thing once, insist on stating it twice, with only minimal difference between the two modes of expression. This tendency is by no means limited to modifiers but may affect nouns and verbs as well as adjectives. Here is a particularly choice example, whose context refers to the dying patient and to recent books on the subject.

The one-time scarcity of writing regarding death and the dying experience *has been supplanted by a profusion of writing from authors of variable credentials, many of whom* blame and discredit *the* physician and his surrogates *for being exquisitely* insensitive and unattentive *to the* social and personal *problems accompanying* fatal and terminal *illness.*

This "doubling," as I like to call it, affects nouns, verbs, and adjectives quite impartially. In each pair there is a slight difference between the two terms. Even in the last pair, *fatal* is not the same as *terminal*, for a patient may have a fatal illness that is not (yet) terminal. But we must ask ourselves, how important are the differences? What does the extra shading contribute? Does a tiny extra shred of meaning compensate for clumsiness and monotony of style? No two words have exactly the same meaning, but can not a single word adequately express the intent of the author?

Actually, doubling is usually a habit. It tends to develop when the author, not quite sure what he wants to say, fails to make his thought precise but instead slings words together in the hope that the messy aggregate will convey his meaning. However, we must distinguish between, say, two adjectives that express quite distinct attributes, or two verbs that indicate different activities, and two that offer only minor variations. Thoughtful revision will get rid of most doubling, and thoughtful editing will eliminate most of the remainder. The residue may prove genuinely useful.

It

It is a pronoun. There are many kinds of pronouns—personal, relative, possessive, interrogative. Each refers to an antecedent, ordinarily a noun (or pronoun), called a *referent.* In the use of pronouns one great source of confusion lies in faulty or unclear reference—the reader is not sure of the antecedent. This difficulty, especially severe with *it*, also affects other pronouns.

In one usage *it* is correlative with *he* and *she*—personal pronouns that distinguish gender. *He* refers to a specific masculine noun, *she* to a feminine noun, and *it* to one that is neuter. In some languages there are only two genders, so that all nouns are either masculine or feminine, while other languages, like German or English, have three. But in English the great majority of common nouns are neuter. If we want to refer to such diverse entities as *book, kidney, gravitation, truth,* the appropriate pronoun is *it* in each instance.

In a second usage *it* has a more indefinite reference. Suppose we say, "It is going to rain." Grammatically, a verb requires a subject. In this example *it* is called the *anticipatory* subject, but the intended subject is the phrase *going to rain. It* "anticipates" this phrase, thus serving a grammatical function made necessary by the peculiarities of English. We see comparable functions in the sentence "It is apparent that he is sick." *It* refers to a deferred subject, namely, the whole clause "that he is sick." This clause is the real subject, and *it* is only anticipatory.

Much of this usage stems from Latin constructions with their impersonal verbs, where the subject, an integral part of the verb form, is not represented by any separate word. Verbs of this character, translated into English, may have a separate subject. For example, *oportet*, "it is proper" (or "appropriate"); *licet*, "it is allowed" (or, to use a different impersonal construction in translation, "one may"); *constat*, "it is well known." To indicate *what* is apparent, or permissible, or well known, the Latin might use a clause in the subjunctive; or, more commonly, have an infinitive construction, whose subject would be in the accusative case. We can get the flavor of this if we think of the English, "It is proper to wear sport clothes," "It is proper for me to wear sport clothes," or "It is proper that I wear sport clothes." The Latin does not need the introductory *it*.

However, there are Latin constructions closer to the English; such as *opus est*, "there is a need" (or "it is necessary"). Here, too, the Latin generally uses the infinitive construction to tell us what is necessary. We see this in the English, "It is necessary to hurry," "There is need to hurry," or more idiomatically, "We must hurry."

In these examples the *it* has a relatively diffuse reference, namely, an entire phrase or clause. For convenience I will call this the indefinite usage, which I would contrast with the highly specific reference wherein the *it* stands for a particular and clearly identified noun. Thus, in the sentences, "Where is my book? Is it on the table?" *it* is a pronoun of unambiguous and specific reference.

Unfortunately, many authors confuse these two, and sprinkle their sentences with numerous *it*s so that the reader loses all track of the intended reference. The single word *it* is forced to do double duty in totally different contexts, sometimes definite, sometimes indefinite. Here is an example.

> It *is a natural impulse, when the manuscript is completed, to put* it *in an envelope and mail* it *to the editor.*

In this sentence the word *it* occurs three times, once as an anticipatory subject and twice as a pronoun referring to specific nouns. But which noun? and How do you know? In the example quoted, the second *it* clearly refers to the *manuscript*, but what does the last *it* refer to, *manuscript* or *envelope?* As a practical matter the reference makes little difference, since the manuscript is in the envelope and if you mail one you are automatically mailing the other. But if we want to be precise, we find scope for confusion.

While there is no rigid grammatical rule, the general tendency is to refer the pronoun to the preceding noun of appropriate gender and number. A singular pronoun will not refer to a plural noun nor the neuter *it* to a feminine noun. But once we have said that, there is a certain amount of leeway. The grammar books tell us that an antecedent should be reasonably close to the pronoun; that no other likely or plausible antecedent should intervene; and, if the antecedent and pronoun are not immediately close together, that the antecedent should be the important word in its context. All this allows considerable possibility of confusion. The reference should always be sufficiently close that the reader need not stop to inquire what the author really means.

Let me give a few further instances. In this example, the subject of discourse is hypertension.

Any proposed concept needs constant testing and revision. It is only in *this way that* it *can be useful.*

The first *it* represents an anticipatory construction, but the second is a relative pronoun whose reference is not immediately apparent. When we look carefully, we see that the second *it* refers to *concept* in the preceding sentence. Intervening between the intended referent and the *it* are the nouns *testing, revision,* and *way.* To find the reference, the reader must skip over these and go back to a noun far removed from the pronoun.

A little care would remove the confusion and also make the sentence much more graceful. Thus, the anticipatory *it* could be deleted with positive gain. The second sentence would then read, "Only in this way can it be useful." But even this is not really clear, for the ambiguous reference of the second *it* still remains. The simplest mode of correction is to recast the two sentences and combine them:

Any proposed concept needs constant testing and revision, which alone can make it *useful.*

The *which* refers unambiguously to *testing and revision*, and the *it* then has a clear track to its referent *concept*. One road to greater clarity is to obey an *ad hoc* rule: Do not have more than one *it* in a sentence.

Here is another example. The discussion concerns a particular television program.

> It *was called* Forum*, and in a rather clumsy way,* it *did attempt to give the public a platform from which* it *could criticize the programs that were being inflicted on* it*.* It *seems to many that* it *would be a good idea to bring* it *back.*

In the first sentence, *it*, occurs four times and the antecedents are by no means clear. I will enumerate them in turn, with their proper referent. The first *it* refers to *program* in the preceding sentence, not given here; the second *it* refers to *Forum*; and the third and fourth, to *public*. Continuing with the next sentence, the fifth and sixth *it*s represent anticipatory subjects; the seventh *it* refers to *Forum* in the preceding sentence.

With a little care we can replace the confusion with a reasonably clear statement:

> *The program called* Forum *tried in a rather clumsy way to provide a platform from which the public could criticize the presentations being inflicted on* it*. I believe* it *would be a good idea to bring back this program, but in an improved form.*

In the first sentence I use *presentations* to avoid repetition of *programs*, which already occurs once in each sentence. *It* also occurs once in each sentence, in the first with the referent *public* and in the second with an anticipatory function.

We could make a further improvement by eliminating the impersonal *it* altogether, substituting a definite subject, and recasting the sentence. For example, "I believe the situation would be improved if we brought back this program, but in an improved form." Or, "I believe we might advantageously bring back this program. . . ." Always try to eliminate an impersonal *it*. "It seems to me that" means *I believe*; "it is perfectly obvious that" means *obviously*; "it is possible that" means *perhaps*. Often we can substitute an adverb for the indefinite construction, and at other times we can simply strike out the whole clause containing the indefinite *it*, without any substitution.

A sentence that begins "While *it* remains true that the most common cause of obesity is simply overeating, nevertheless . . ." can be markedly improved by a deletion: "While the most common cause of obesity is simply overeating, nevertheless . . ." The clause "it remains true that" adds less than nothing. The anticipatory *it* is usually easy to eliminate if only we think to do so.

When we worry about the reference of *it*, we must also worry about the reference of other pronouns. Thus,

> *Tractors are not seen nor used, for the people cannot afford* them, *and if* they *could,* they *would be utterly useless in the terrain.*

In this sentence, *them* raises no problems, for it clearly refers to *tractors.* But the two *theys* are confusing: The author wants the first *they* to refer to *people*, and the second to *tractors* at the beginning of the sentence, but this is not what the sentence actually says. Some change is imperative. I suggest this alternative:

> *Tractors are not seen nor used, for the people cannot afford* them, *and in any case* they *would be utterly useless in the terrain.*

Them refers to *tractors*, and the *they* refers to *them* and thus indirectly to *tractors.* There is no ambiguity.

Here is a splendid example of confused reference:

> *This represented a challenge to our mature values* which *had to be eliminated as soon as possible.*

As written, the sentence declares that the *values* must be eliminated, whereas the author intended that the *challenge* should be disposed of. We can bring out the intended reference by a suitable modification:

> *This challenge to our mature values must be eliminated as soon as possible.*

66 A somewhat more complicated example, and one more difficult

to correct, comes from a medical case report. The patient was showing progressive improvement:

His temperature became normal but he contracted a urinary tract infection secondary to an indwelling catheter on the tenth hospital day, which responded to antibiotics.

What was it that responded to the antibiotics? the hospital *day*— the immediate antecedent— or the *catheter* or the *infection* or the *temperature*? The sense demands that the referent should be the *infection* or possibly the *temperature*, and to bring this out the sentence must be massively revised. I suggest this version:

His temperature became normal but on the tenth hospital day, because of an indwelling catheter, he contracted a urinary tract infection which responded to antibiotics.

Whether the *which* should more properly be *that* is a problem I will not discuss here.

One further example:

Discussions following each paper bring out differences of opinion on controversial subjects which unify the text into a vigorous and informative treatise.

To what does the *which* refer? The immediately preceding noun is *subjects*, but do the subjects unify anything? Such an interpretation makes no sense; we must seek further. There are several other antecedent nouns, but since *unify* is a plural form, *which* must be plural, and therefore its referent must be plural. This requirement eliminates *opinion* or *paper* and leaves either *differences* or *discussions* as the intended term. After a moment of reflection we realize that it is the *discussions* that unify the text. On the principle of placing together the terms that belong together, we might transpose and alter the original sentence into

Discussions, which unify the text into a vigorous and informative treatise, follow each paper and bring out differences of opinion on controversial subjects.

67

But this is rather awkward. To effect a more meaningful improvement we should first try to puzzle out what the author is trying to say. He is talking about the *discussions*, and he points out that these discussions accomplish two things—they bring out differences of opinion and they also unify the text. Why not say so, simply and clearly? All we need do is delete the *which* in the original sentence and substitute *and*.

> *Discussions following each paper bring out differences of opinion on controversial subjects* and *unify the text into a vigorous and informative treatise.*

We now have a simple sentence with a single subject and a compound verb. The ambiguity has disappeared, and there is no stumbling over the meaning.

These last examples indicate some of the difficulties that attend the proper use of relative pronouns in general. Our concern about the reference of *it* should extend to the reference of all pronouns. Each should be tested in the same way as *it*, to make certain that its reference is quite unambiguous.

References

1. Asher, R.   Six honest serving men for medical writers. *J.A.M.A.* 206:83, 1969.
2. Carlyle T.   *Sartor Resartus.* London: Dent (Everyman's Library), 1973. P. 127.
3. Disraeli, B.   *Coningsby.* London: Dent (Everyman's Library), 1971. P. 291, 396.
4. Rosenberg, C. E.   *The Cholera Years.* Chicago: University of Chicago Press, 1962. P. 147.

Starting to Write

Modes of Writing

There are as many different ways of writing as there are writers. Some novelists—especially detective story writers—prepare a rough plot for the first few chapters, sit down to write, never to look back, and let the story and the characters develop themselves. In most cases the results are sloppy in the extreme, but there are many exceptions. Rex Stout, for example, one of my favorite authors, told a newspaper reporter that he never revised his work. He never rewrote. His stories, he said, took form while he was writing them, and he sent his typescripts directly to the publisher without revision [2]. The only rereading he did was the output of the preceding day. Georges Simenon, I believe, had a comparable method of working. They exemplify writers of vast talent, disciplined imagination, and fluent writing technique.

A newspaper reporter may write as rapidly as his fingers can pound the typewriter and send off his copy as fast it is written. Samuel Johnson wrote *Rasselas* in enormous haste, giving the handwritten sheets, as fast as he finished them, to a printer's boy who stood waiting at his elbow. Rudolph Virchow, we are told, did not revise. His manuscripts, where still extant, show negligible corrections.

The noted medical historian Henry Sigerist gave a detailed account of his own method of writing [3]. He had, apparently, a remarkably systematic mind. When starting to write a major work, he would first prepare a brief outline, approximately a page in length. This would suffice for a journal article or a single chapter in a book. Every working day he would write for three hours, from 9:00 in the morning until noon, and during this period he would finish about five pages, some 700 words in all. Using special copy books of folio size, he wrote in longhand on the recto only, while the facing left-hand page he kept for footnotes and minor changes. But the changes would be minor indeed; for practical purposes what he wrote was in absolutely final form.

Such a mode of writing, which seems so straightforward, actually represents only the visible tip of the iceberg. A vast amount of preliminary work underlay the writing of the text. In a different metaphor, there had already been a long gestation period. Sigerist himself declared that, when he decided to write an article or a book, he felt "pregnant" for a long time. There had been much reading, 71 much note-taking, much reflection. But for the most part he ordered his thoughts in his head. When the time came to commit these thoughts to paper, the work was already mostly done. Such a mode of composition requires an especially methodical mind that is rare indeed.

Other writers work in an entirely different fashion. As a foil to Rex Stout, I would mention P. G. Wodehouse, whose writing flows so smoothly and whose understatements and delicate incongruities are such a delight. In an interview he said that he rewrote every page nine or ten times [4]. Yet the reader is never aware of the labor involved.

Rachel Carson was similarly a perfectionist. If we assume the initial talent, then writing, she declared, is "largely a matter of application and hard work, of writing and rewriting endlessly until you are satisfied that you have said what you want to say as clearly and simply as possible. For me that usually means many, many revisions" [1].

There are many variables involved. Some authors "naturally" write well; others have little natural skill. Some put down what they have to say and never revise; others make innumerable revisions. Some have a drive to achieve perfection and can always find room for improvement; others are easily satisfied. And there are innumerable gradations. In this book I direct myself to those who are not satisfied with the way they write and who want suggestions for improvement; who realize that, for all except the rarest genius, good writing means hard work, frustration, and anguish; who feel that the effort at improvement is worthwhile, that craftsmanship, in whatever discipline, is worth achieving, and that enormous satisfaction results from work well done.

Types of Expository Writing

In this book I deal only with expository writing, that is, writing whose function is to describe, explain, interpret, and make clear. Exposition rests on a basis of facts, but mere presentation of facts is not enough. A telephone directory, for example, offers a massive collection of facts, suitably arranged, but it is only a compilation, not an exposition. There is no synthesis, no forward reference, no conclusion. So, too, with a table of logarithms. The table itself is a compilation, but the set of directions that tell you how to use the logarithmic table would be an exposition.

We may think of all exposition as existing on a scale ranging from objective to subjective, depending on the degree to which the author intrudes himself into the writing. A simple objective exposition is a set of directions, such as the instruction sheet that tells how to put together a toy or an apparatus, how to proceed from the separate elements in isolation to the finished product. Exposition can vary not only in objectivity but in clarity, as every parent knows who has tried to assemble toys on Christmas Eve: "Place the thule inside the grimble and after inserting the gismo, tighten the slobe around the dreeble"—or equivalent. This describes but does not clarify. As exposition it is not very good.

In directions of the "how to do it" type, there is a minimal subjective component and no expression of opinion. At the opposite pole we might place, say, the editorial which lays special emphasis on the author's opinion on some specific topic, whether politics, sport, economics, social science, history, or whatever. In the editorial, or any frankly subjective exposition, we have the background of facts, which I would call the data base, and a large amount of admittedly subjective interpretation and comment.

Most exposition falls somewhere between these two extremes. In the sciences—whether social, biological, or physical—an author might provide a review of the literature, describing the findings of others; or give an account of original research; or promulgate a new theory. All these are types of exposition. A textbook is a long and carefully ordered exposition; an encyclopedia is a collection of expository essays. Most exposition, combining presentation of data with interpretation, demands critical judgment and evaluation. The author gives his opinions and his reasoning in addition to the supporting evidence. Expository writing, then, conveys information mingled with varying degrees of interpretation and opinion.

The Preparatory Work

All expository writing demands preparation, the nature and duration of which depend on the character of the exposition itself. What are you planning to write? a 350-word book review, a case report for the medical journal, a 5,000-word chapter in a book, a

10,000-word term paper, or a 100,000-word doctoral dissertation? In any case, you collect your material—the data with which you work—you reflect, and you write. If you are writing a book review, you must first read the book, then reflect. If you are going to write a case report, you study the hospital record, read the relevant literature, and then reflect. If you are writing history or biography, you hunt out the sources, and then reflect. For most of us all this requires note-taking. You make abstracts of your reading, or quotations therefrom; you record the thoughts that come to you as you do your reading—the critical judgments, the points of agreement or disagreement; you note the ideas that need further elaboration. The whole range of mental processes that we call associations, I summarize as "reflections." These should be written down, for if they escape you at the moment they may never be recaptured. Notes can always be discarded if they prove useless; ideas, once gone, may never return.

Reading and reflection involve a circular or, better, a spiral process. Reading leads to reflection, which in turn leads to more reading, which in turn leads to further reflection and new associations. Each turn of the spiral modifies what has gone before. You read over your old notes and you get new ideas that you had not had when your knowledge of the sources was still embryonic. Reflection on the sources can set off an associative reaction, strike a response, lead to some sort of interpretation, fugitive unless recorded.

I would strongly recommend note-taking (and notations of ideas) not on cards but on uniform-sized scratch paper. I use typewriter paper cut in half. With this you can be utterly profligate without worrying about excessive bulk or cost. Place your notes, thoughts, quotations, queries, and lists of agenda, divided according to topics, in envelopes of appropriate size, suitably identified and filed. Have your scratch pads (or loose sheets on clip boards) constantly available, since useful thoughts can come to you at the oddest times. But in addition to seizing the random thought as it passes by, you must deliberately sit down to think, going over masses of notes, letting them speak to you, and recording what they tell you. Perhaps after a few such thought sessions you will decide to rearrange your approach, discard certain aspects as not relevant, identifying areas that need further data, and generally revise your concepts.

If you are writing something quite short you can soon collect your notes and start to write. Short compositions can be finished in a single session, and a first draft dashed off in a single burst (subject to rewriting). But if you are engaged on a major project, you should not try to finish your preparation before you begin the task of writing. I would strongly recommend writing preliminary segments, perhaps 200 to 500 words at a time, not in any systematic fashion but as the ideas well up in your mind. These segments are, in a sense, formalized notes. The discipline of putting them down in an orderly fashion provides a structure and coherence that scribbled notes do not have. Such segments may or may not appear in your final draft. Most likely they will find a place, somewhere, but with suitable modifications. Meanwhile you have made your thoughts orderly and ready for revision.

What Are You Trying to Say?

Expository writing involves the problem, What are you really trying to say? a question that on its face seems easy to answer—all you need do is point to the finished work, the 10 or 20 or 100 pages that you have just written. But if this were the answer, then editors and readers would have a thoroughly easy life. For most writers, what they are trying to say and what they actually have said are quite distinct, and all too often what they have said is not at all clear to the reader and not really to the writers themselves. The reasons are many, but fortunately are subject to analysis.

The story is told of an experienced author who had already completed a large part of his manuscript. When asked what he was trying to say, he replied, "How do I know what I want to say until my typewriter tells me?" This answer is worth pondering. It implies that, no matter how precise the intention or the original outline, an author is never sure of what he wants to say until he actually gets it down on paper. Even then he is not really sure, but has at least a good inkling, which will become clearer with each rewriting. Regardless of preliminary thought, there is a constant flow of new ideas down through the fingers into the pen or onto the typewriter keys. As we write, new associations appear that did not come to the surface until the writing process started. Often previously unsuspected categories and topics come to mind, as well as many details whose relevance had not previously been suspected. But all this takes place only as the original thoughts get transferred to paper.

We have probably all heard of the novelist who may plan his novel with a definite plot and characterizations, but who finds as he writes that the novel seems to take on a life of its own and that the characters want to do things that he had not originally intended. Rachel Carson dealt with this phenomenon. In reference to her own *The Sea Around Us*, she wrote that the author must not impose himself on his subject. His task, she declared, is to know the subject intimately, to let it fill his mind. Then, at some point, "the subject takes command and the true act of creation begins. . . . The discipline of the writer is to learn to be still and listen to what his subject has to tell him" [1].

Of course, the psychologists speak of the subconscious, of the intellectual activity that goes on without our awareness and comes into consciousness at some future time, often without warning. Unlike Sigerist I find that new ideas are constantly emerging every time I start to write. This is one reason why, when still in the preliminary stage, I like to write in short segments. Something analogous to a fermentative process goes on in the subconscious, whereby "bubbles" continue to rise to the surface over a long period of time. Eventually there comes a time when new thoughts no longer bubble up freely, and that is the moment to start the second phase of expository writing—revision.

I maintain that an author cannot answer the question, What are you really trying to say? until he has made a good start on his work, and has permitted whatever exists on paper to interact with what is still in his head—thus favoring the fermentative process going on in the subconscious.

An important corollary emerges: If we expect an interaction between early written drafts and thoughts that are not yet conscious, then we cannot rigidly plan ahead of time what we are going to say. In more concrete terms this means that we cannot adhere slavishly to an outline. Many books on writing recommend drawing up a meticulous outline before starting to write. These books recommend doing all the preliminary work—all the necessary reading—then preparing the outline and arranging all notes in logical fashion.

Then, according to this advice, the work is chiefly done, and all that remains is the actual writing, or, as I would say, putting flesh on the skeleton now so neatly articulated. According to this view the process of writing becomes simplified and can receive the author's undivided attention. He can focus on good grammar and clear expression rather than on content (which is all arranged).

With this viewpoint I must disagree. Some sort of outline is of course essential so that the writer will not lose his way completely. But the process of doing all the preparation first, then placing the thoughts in logical order, and then doing all the writing, may work for some people but I cannot recommend it. The thinking and the writing cannot be sundered, and the notion of preparing a detailed skeleton outline, and then letting that outline determine the final shape of the work, is rarely satisfactory.

We can compare the writing of an essay or a book with the process of gestation, as an embryo develops from its first germ to the full-term baby. An animal does not come into being with a skeleton as the first step. Nature does not first create a skeleton and then add layer upon layer of flesh. Instead, from the moment of conception, an animal is an organic whole that grows, and as it grows, differentiates. If we regard an outline as a skeleton to be fleshed out through elaboration of notes arranged in a definite order, then we have a fine procedure for a taxidermist. But the product of a taxidermist is rather lacking in vitality. A writer should create his brainchild along the lines of natural embryogenesis, with progressive differentiation, wherein the different parts react on each other and affect each other as they develop. Of course, he must have a general plan of growth. We do not want a human embryo to turn into an ape somewhere along the line, and if we are breeding rabbits we do not want to end up with guinea pigs. But there is a vast difference between growth according to a plan that exhibits plasticity, and growth that accrues on a rigid skeleton.

In one sense the problem is verbal: an outline can be flexible, or it can be rigid. I am violently opposed to the latter and to those books that recommend it. If we construe "outline" as something flexible, then I raise no objection, but, to avoid ambiguity, would prefer to say "plan" rather than "outline."

However, in another sense the problem is much more than verbal, for it relates to the very process of creative writing itself. I have two chief objections to writing strictly from an outline. First, it tends to stifle that major aspect of creative writing, revision, or at least to inhibit all but minor changes. And second, the rigid outline, if followed, deflects any benefits from later ideas, fails to nurture them, prevents reaping any profit from maturing thought. By way of example, I would point to the experience I am sure all of us have had: When reading a book of some difficulty, we may underscore or otherwise mark certain passages; if we return to that book after a lapse of time, we may wonder, "Why did I ever mark that, when *this* is so much more significant?" Greater familiarity with the subject has brought new insights, so that what formerly had passed us by now reveals its full importance; and, conversely, what had once seemed central is now recognized as only tangential. The same process occurs with note-taking. If we go back to the original texts, we may find that our notes missed certain crucial features whose importance we had not recognized at first.

Perhaps all this would come under the category of feedback, a constant interaction between the new and the old, each appropriately modifying the other. Good expository writing demands such feedback, or at least demands scope for its occurrence. Hewing to a rigid outline denies this scope. I would urge flexibility at all points, from the moment that the idea of writing first arises until final completion of the work.

In the writing process there are many problems and many ways of meeting them. Indeed, so numerous are the difficulties that we should rather speak of categories or clusters of problems. I suggest three interrelated categories. One has to do with knowing what you really want to say; the second, with getting it all down on paper; and the third, with trying to say it better. In my view, knowing what you want to say is a process that comes through reading, reflection, and actual writing. Reflection is reinforced through writing; the writing itself, and the thought that emerges therefrom, are in turn accentuated through revision. And then the whole interaction may send you back to do more reading or rereading. Here then are four variables, reading, reflection, writing, and revision. (Difficulties with an outline are included in "reflections.") Each affects all the others; each depends to some extent on the others.

In this book I want to offer some suggestions not for reading or reflection—any proper discussion of these would require another whole book—but for the actual writing and for revision. Revision and editing are closely allied. If you edit your own work you are engaged in revision. If you revise someone else's work, you are engaged in editing. For didactic purposes it is much easier to revise someone else's work than your own. In the next two chapters I will take up the problems of revision and editing and of bringing your own writing to a conclusion.

References

1. Brooks, P. *The House of Life: Rachel Carson at Work.* Boston: Houghton-Mifflin, 1972. P. 1–3.
2. *Chicago Daily News,* September 30, 1967.
3. Sigerist, H. E. Thoughts on the physician's writing and reading. In *Medical Writing* M.D. International Symposia #2. New York: M.D. Publications, 1955. P. 3.
4. *Time,* February 24, 1976.

5

Editing and Revising

Seeking the Aesthetic Fit

Revision has to do with change, specifically with finding the particular change that will bring about improvement. The motive force is an inner dissatisfaction, the vague feeling that something is not good enough. You try one change after another until that dissatisfaction is allayed, until that inner restlessness gives way to a sense of rest, at least, temporarily. The satisfaction may be only temporary, and unrest may flare up again at a future time.

To illustrate this I offer the example of arranging furniture when you move into a new house. Where should the sofa go? What relation should it bear to this chair, or that table? You try it in one position and you are not satisfied, so you move the sofa a little to the left, or perhaps to the right. Then you go across the room and look at the result. And perhaps move the sofa back a little. And the next day you may decide that you don't like it at all, and make a totally different arrangement, putting the sofa on the other side of the room altogether. You may continue the rearrangement until you finally reach a disposition that seems just right. You say to yourself, "This is *it*." There is a gut feeling, the same aesthetic reaction that an artist experiences when he arranges (and rearranges) a composition until he is satisfied. The artist Tom Wesselman expressed this effectively. He declared that he arranged and rearranged things until the elements "lock" into place. When this happens, he said, the composition seems to develop a physical resistance to further manipulation [2].

I call this an aesthetic satisfaction, or, in more popular vocabulary, a sense of fit. The artist has such a sense highly developed in regard to form and color; the musician, in relation to tone, harmony, and rhythm. Other persons, less well endowed by nature, who cannot carry a tune or keep in rhythm, or who are not sensitive to color combinations and spatial balance, have a poorly developed sense of musical or artistic fit.

Sensitivity to language I regard as exactly similar. Some persons take delight in the felicities of language—the well-turned phrase, the clear statement, the precise discrimination, the graceful construction. Other people are quite indifferent to such nuances. However, just as with music or art, sensitivity can increase through practice, if adequate motivation will fuel the necessary effort.

Various techniques for achieving improvement I take up in other chapters.

In seeking improvement, you must first of all find a way of saying things differently, and then decide whether the new way is better than the old. Whoever wants to achieve a feeling for language must have a constant flow of alternatives. *How else can I say this?* should be the query always present in the back of your mind. The more alternatives you can dredge up, the broader your field of choice. These alternatives may flow along two different channels, one having to do with words, the other with grammatical constructions.

A given word may impress you as not quite right, not sufficiently precise. Can you find another word, with a meaning not identical but at least overlapping and closely related? Such a synonym may fit better, may express more satisfactorily the lurking thought. If one synonym does not quite meet the need, then perhaps another will. The well-stocked mind, like a well-filled reservoir, will provide a constant flow of words. Open the linguistic faucet and half a dozen words come out. Savor each in turn until you get the one that provides the aesthetic satisfaction or sense of fit. And if the precise, aesthetically satisfying word does not come to mind you can have recourse to a thesaurus, such as Roget's. The thesaurus, however, does not tell you which word is best, but merely supplies a range of alternatives among which you must make the choice.

Having a reservoir of alternative words is one resource to be cultivated; equally important is a store of alternative constructions. You can put words together in more than one way. To express a given thought should you use the active voice or the passive? a subordinate clause or a prepositional phrase? a short sentence or a long one? Is it better to put a given word or phrase in this position or in that? Questions of this character should constantly go through the mind of an author, not perhaps in the initial writing but as he revises.

However few or many the answers that occur to him, the author must choose the one he considers preferable, and he should be able to defend his choice. If an author is dissatisfied with a given construction and cannot think of a ready alternative, he cannot go to a book for help. He is entirely on his own. If his reservoir of alternatives is low, he may sweat and agonize until something does come to mind. He may wrestle with various alternatives. Sooner or later he will hit upon a likely possibility that may satisfy him for the moment—or at least until the next revision. Through experience an author will gradually fill up a reservoir of alternatives that he can readily draw on. In this way he builds a technical proficiency. Verbal technique and aesthetic sensitivity should develop hand in hand.

The process of revision and the process of editing have much in common. They both aim at improvement, but the one you direct toward your own work, the other toward someone else's work. Ordinarily, it is much easier to see the faults in another, whereas your own faults slither away from your critical vision. Toward someone else you can be objective, but it is much easier to perceive the mote in your brother's eye than the beam in your own.

Let us start with some problems of editing, that is, of bringing improvement into the writing of someone else. After you have had even a modest experience with work of this kind, after you have seen how a relatively small change here and there can make the text sound better, you can more easily approach your own work, see faults that might otherwise have escaped your attention, and readily find ways to correct them.

In my writing classes I provide samples from "live" manuscripts, which I instruct the students to edit as they see fit. At first the possession of an editorial pencil seems to transform a mild, earnest student into a tyrant who first slashes the text to bits and then recasts these bits into a totally different form. When I ask the reason for this, the student will say something to the effect that the new rendition "sounds better." Actually, this may or may not be true but is not entirely relevant to the process of editing. I ask the student, What will the author think when he sees your corrections? Will he recognize the work as his own? After all, it is he who signs it. It is his brainchild and he should be able to recognize his own parenthood. I then expound my credo in regard to editing: Preserve the original words and constructions as much as you can, so that the author is never in doubt that *he* has written it; conduct your editing so that the author, when he sees his article in print, will think, "I did not realize that I wrote so well."

Making Minor Changes

We should regard a manuscript as a patient who may need surgery. The procedure may be either minor or major, depending on the condition. The editor-surgeon should make only those changes absolutely necessary to render the prose acceptable, and he should deliberately try to keep his alterations to a minimum. Usually a minor operation will suffice.

My concept of minor surgery involves three components: simple deletion; simple substitution; and transposition of a word, phrase, or clause. Let me give some examples.

The first is the simplest possible case, deletion of a few unnecessary words. The context has to do with evidence and the way a physician reaches a decision.

> *The physician constantly deals with evidence. The clinician in practice rests his diagnosis on the basis of evidence that he calls "signs and symptoms"; the medical scientist builds his theories on the basis of evidence drawn from his observations and experiments.*

This does not cry aloud for improvement, but it does contain unnecessary words. "Clinician in practice" is tautological. If we delete *in practice*, we eliminate the redundancy and sharpen the contrast between *clinician* and *medical scientist*. Then, the words *the basis of* are redundant; they detract rather than add. With simple deletion of these redundancies the sentence reads,

> *The physician constantly deals with evidence. The clinician rests his diagnosis on evidence that he calls "signs and symptoms"; the medical scientist builds his theories on evidence drawn from his observations and experiments.*

Although the word *evidence* occurs three times, the repetition I regard as deliberate. It adds a desirable emphasis.

The next example also shows unnecessary words—useless verbiage. The cure is a simple excision.

> *With increasing demands for chemical information on every aspect of all kinds of environmental hazards as it affects man and his environs,*

the chemical sciences must have a broad, multidisciplinary, and active role in providing man and his world a safer environment than presently now exists.

The need for deletion is obvious. After excision, the sentence reads,

With increasing demands for chemical information on every aspect of environmental hazards, the chemical sciences must have a broad, multidisciplinary, and active role in providing a safer environment.

If I were writing this text, I certainly would not have expressed my thoughts in this way. But I was not the author. My task as editor is to help the author say what he wants to say, and to say it better than he did originally. The goal is improvement, not perfection.

We can distinguish two types of redundancy. In one, the excess words merely say again what has already been said in some other way—a covert repetition. In the second, there are words whose presence adds nothing, whose absence detracts nothing. They merely take up space. The next example illustrates both types. The cure involves not only deletion but also a minor transposition.

This small book edited by Dr. John Doe has as contributors eight highly experienced physicians who have had a great deal of clinical experience in the treatment of blood disease in their respective special areas of expertise. In addition to their clinical experience, they have participated in both basic and clinical research in the subject.

There is much unnecessary repetition here. The clause "who have had a great deal of clinical experience . . . in their respective special areas of expertise" merely repeats what is concisely expressed in *highly experienced*. If we want to know wherein the experience lies, we find the answer in the phrase *in the treatment of blood diseases*. Deleting the repetitive terms in the first sentence, we have

This small book edited by Dr. John Doe has as contributors eight highly experienced physicians in the treatment of blood disease.

However, this would sound much better if we make a small transposition, so that the words that belong closely together are in fact together. *Physicians* should go directly next to *eight*, and *experi-*   85 *enced* next to *in the treatment of blood disease*. Thus,

> *. . . eight physicians highly experienced in the treatment . . .*

is a definite improvement.

In the second sentence of the paragraph the phrase *in the subject* adds nothing and its deletion would automatically be an improvement. When all these changes are made, the whole passage reads

> *This small book edited by Dr. John Doe has as contributors eight physicians highly experienced in the treatment of blood disease. In addition to their clinical experience, they have participated in both basic and clinical research.*

Again, the editor had the task of improving the original, not of achieving perfection.

The next two examples show the need for all three components of minor surgery—deletion, substitution, and transposition.

> *In the preface we learn that the contributors were asked to provide information for use to [sic] the primary care physician as he labors to better understand and treat disease. In addition the book provides the specialist in pediatrics, endocrinology, or metabolism a useful reference source of related fields.*

In Chapter 3, I recommended eliminating introductory clauses ending in a *that*. The very first clause in this example, "In the preface we learn that," should go. To be sure, we lose thereby the information that the particular statements occur in the preface (instead of in a numbered chapter) but the value of such information is negligible. A grain of wheat does not compensate for a bushel of chaff.

The first sentence shows not only the redundancy of an introductory clause, but also a barbarism, *for use to*, and a clause that is stilted and precious, "as he labors to better understand." Some simpler expression would be preferable. I suggest editing the first sentence to read

> *The contributors were asked to provide information for the primary care physician to help him better understand and treat disease.*

"To help him better understand" does not have the identical sense of the original, but the added compactness more than compensates for the minute change in sense.

The second sentence contains both a direct and an indirect object. What does the book provide? It provides a *reference source*. For whom? For the specialist in the various fields enumerated. In the original version the preposition *for* is implied. To improve the sentence we should first of all place verb and direct object close together: "the book provides a useful reference source." Then where will we put the indirect object, *for the specialist*? It also belongs close to the verb. The solution: Place the indirect object before the verb and the direct object immediately following:

. . . for the specialist . . . the book provides a useful reference. . . .

The whole passage then reads

The contributors were asked to provide information for the primary care physician to help him better understand and treat disease. In addition, for the specialist in pediatrics, endocrinology, or metabolism, the book provides a useful reference source of related fields.

The next example is a little more complex.

As in the previous editions this text is directed to and intended for those physicians who devote the main part of their professional lives to the care of children, be the physician a pediatrician, general practitioner, school physician or other.

As noted in Chapter 3 I dislike a construction wherein two prepositions govern a single object. It is not ungrammatical, but since to me it sounds awkward, I try to eliminate that construction when possible. Here the solution is simple in the extreme, for the pair *directed to and intended for* is repetitious. We need merely delete one of the pair: "this text is intended for those physicians."

The rest of the original sentence describes the kind of physician for whom the text is intended, but does so in wordy fashion. The clause "who devote the main part of their professional lives to the care of'' can be shortened to ''who devote themselves mainly to the care of.''

Then, the repetition of *physician* is extremely awkward. As remedy delete the second occurrence and transpose the enumerated classes to be in apposition with the first *physicians*. This transposition requires a few minor changes in grammar. The whole sentence would then read

> *As in the previous editions, this text is intended for those physicians—pediatricians, general practitioners, school physicians, or others—who devote themselves mainly to the care of children.*

The last example of what I call minor surgery illustrates the need for simple deletions and simple substitutions.

> *An almost inconceivable quantity of investigation, over the past several decades, finally has provided insight into some hithertofore mysteries of many crippling degenerative diseases. The information, admirably documented in this work, conveys at least a thread of hope that with this impetus, continued efforts will eventually produce effective therapy for the unfortunate victims of a multitude of degenerative diseases.*

Many of the words and phrases, utterly superfluous, we can delete with advantage. Then, the author seems to have adopted the policy, Why use a short word when you can use a long one? We should try to find simpler ways of conveying the ideas. For example, instead of saying *An almost inconceivable quantity of investigation*, I would suggest *A vast amount of study*, although other alternatives are, of course, possible. I would make deletions and substitutions so that the passage would read:

> *A vast amount of study, over the past several decades, finally has provided insight into many crippling degenerative diseases. The information, admirably documented, provides at least a thread of hope that continued efforts will eventually produce effective therapy.*

Major Alterations

Sometimes a manuscript or a portion thereof will require changes so extensive that they could scarcely be called minor. The editor, instead of making a few deletions, simple substitutions, or transpositions, will need to recast the passage completely. Then the result will no longer be recognizable as the work of the original author. This may be unfortunate, but so too is the amputation of a leg if the well-being of the patient demands it. In line with our metaphor, we can call massive textual changes "major surgery." Let me give a few examples. In the first the context has to do with medical education.

> *Within the medical specialties compartmentalized in named departments, there are "core areas" which need to be taught and within which experience by practice is necessary for adequate mastery, no matter whether the product mounts the academic ladder or contentedly exists as a dispenser of services in city or county, singly or in groups.*

After several readings the sense of this will become clear. But an editor, among his various functions, should promote ease of comprehension and simplify the task of the reader. A passage like the above should be handled ruthlessly. The last half of the sentence says nothing and can be eliminated. The first half has a message that can be rendered quite simply:

> *Every medical specialty has a core of information that must be taught, but for adequate mastery practical experience is necessary.*

The author will not recognize this as his own style, but he may then realize that his style desperately needs improvement.

The second example has to do with epidemiology and the spread of infection in a community.

> *The epidemiology of parasitic infections in human populations is closely related to a number of bionomic and ecological factors which affect the survival and propagation of the causative pathogenic agents. Concomitantly, it is also greatly influenced by social and economic conditions of the population, the state of hygiene and sanitation in the community and the patterns of work and behavior of its members.*

There is nothing here that we can call wrong. The sentences are entirely grammatical and the individual words all parse. And yet, despite the technically correct grammar, the writing is heavy and opaque. It reminds me of a soggy biscuit. If we analyze the passage we find an overabundance of words and an overabundance of syllables. The reader gets tired after a single paragraph, and if he had to go on for page after page, he would get very tired. If we wanted to make the style simpler and more appealing, what could we do? We cannot merely cut a word here and there. We would need to go over the passage with minute care, to find a short word for a long one, to cut out unnecessary words—to put the biscuit in the oven and dry it out. The result, however, would no longer be recognizable as the style of the author. We might, for example, come up with something like this:

> *The spread of parasitic infection in man depends on many biologic factors that affect the causal agent. Also important are various host factors— social and economic conditions, hygiene and sanitation in the community, and the patterns of work and behavior.*

In this rendition I have inserted the words *host factors* to balance *causal agent*.

This degree of editing, although appropriate for short passages, is not feasible for an entire manuscript, nor is it even desirable. As a matter of practical editorial management, if I were handling this manuscript, I would return it to the author for shortening; to help him, I would revise two or three paragraphs, as models, and suggest that he perform the same type of surgery on the remainder.

Here are further examples that require drastic changes, although perhaps less severe than the preceding ones.

> *Despite this, it is evident that psychological stresses do contribute to the precipitation of allergy attacks and to the aggravation and persistence of clinical symptoms.*

By this time, *Despite this, it is evident that* should grate on the reader's ear. For these six words we can substitute the single word *Nevertheless.* For the rest of the sentence we should substitute verb forms for prepositional phrases. The passage will then read,

> *Nevertheless, psychological stresses do precipitate allergy attacks and aggravate clinical symptoms.*

The suitability of *allergy attacks*—the noun-modifier—I will not discuss here.

As a further example I offer a quotation that raises instructive problems. The author has compressed into a single sentence two distinct thoughts that have become jumbled through unskillful writing.

> *In the past, physicians have rarely evaluated and treated patients suffering from lower esophageal disease with understanding, confidence, and success.*

The first problem is, What does the phrase *with understanding, confidence, and success* modify? Not the immediately preceding *disease*, but the more distant verb which, however, is compound—*evaluated and treated*. Does the phrase modify both components equally, or only one, or partly one and partly the other? Does it make good sense to speak of evaluating patients with understanding, confidence, and success? Does it make good sense to speak of treating patients with understanding, confidence, and success? I would say no. I believe we evaluate with understanding and treat with confidence (and, if lucky, with success). If this is the correct interpretation, then we must recast the sentence to make the point unequivocally clear. This will require a considerable rearrangement. Thus,

> *When patients suffered from lower esophageal disease in the past, physicians rarely evaluated them with understanding or treated them with confidence and success.*

I believe this is what the author wanted to say.

As a final example I present a passage from a book review.

> *The authors state that the objectives of this text are two-fold. First, the book is to demonstrate that nutrition has become a clinical science which, despite its complexity, possesses a logical orderliness based on physiology. Second, the book is not intended to be a reference work but rather a formulation of clinical diagnosis and management founded on physiological principles.*

First we delete *The authors state that*. Then we note an intent to provide a parallelism, indicated by a *first* and a *second*. However, the parallelism, undoubtedly clear in the author's mind, grammatically is a little clouded. *The book is to demonstrate* and *The book is not intended to be* do not quite match, and in any case *The book is to demonstrate* is a barbarism. We thus have the problem, How can we preserve the parallelism in the clearest and simplest fashion? We do so by substantial deletions and some changes in construction.

> *The objectives are two-fold: first, to demonstrate that nutrition has become a clinical science based on physiology; and second, to provide a formulation of clinical diagnosis and management founded on physiologic principles.*

The material deleted does indeed have some relevance, but in my opinion only to a trivial degree. That science has a logical order, is scarcely news; that the authors did not intend to write a reference text is perhaps slightly more important. But if we insert this we detract from the impact of the parallelism. The revision, I believe, is more effective as it stands, without further qualification.

Editorial Niceties

Editing demands a critical attention, a mind constantly alert to discrepancies, to clumsy expressions, to rough spots that interrupt the flow of the prose. The editor must regard the sense of his text and at the same time the mode of expression. He must recognize the passages where the sense becomes obscure or the presentation needs improvement.

The alert editor will, for example, pick up the inadvertent repetition of a word. The author, concentrating on what he is trying to say, may not notice such a repetition. To the editorial eye, not clouded by the original pressure of thought, a word used, say, five times in five successive sentences, will stand out sharply. The editor will then find some way to get rid of the repetition—perhaps by introducing synonyms, or perhaps by recasting the sentence.

Excessive use of a given word is one instance of an "obvious" fault that nevertheless has escaped the author's attention. Another fault that can easily escape the author but that an editor will readily perceive is the dangle—the word or words that sit in isolation, modifying nothing, attached to nothing. Most common is the dangling participle:

Not possessing a copy of the manuscript, the basis of my translation was the third edition of 1742.

The participle *possessing* does not modify anything. It has no grammatical relationship with the rest of the sentence. The intent is clear: Because *I* did not possess the manuscript, it was *I* who had to use the third edition. The sentence as given, however, does not say this, even though the implication is there. We make the sentence grammatical by saying

Because I did not possess a copy of the manuscript, the basis of my translation was the third edition of 1742.

Similarly with the sentence

Exploring this possibility, sternal punctures were performed on two other patients.

As it stands, the participle seems to modify the noun *punctures*, which is absurd. The author intended to say,

Exploring this possibility [or better, *To explore this possibility*] *I performed sternal punctures on two other patients.*

Note that dangling participles can occur more readily with the passive voice.

In both these instances the author's thoughts ran ahead of his fingers. There was a compression of ideas, the grammar was fractured, and some participial splinters were left over.

Nouns can also dangle.

An excellent teacher and investigator, his influence spread over all western America.

93

Teacher and investigator stands isolated, serving no grammatical function. Again the intent is clear. The nouns were to be in apposition with someone (not expressly stated) who spread the influence. The correction may take different forms. We can provide a *he* as the real subject.

> *An excellent teacher and investigator, he spread his influence over all western America.*

Or we can provide a verb for the isolated and dangling nouns,

> *Since he was an excellent teacher and investigator, his influence spread all over western America.*

An adverb may also dangle, as with the popular usage of *hopefully:* "Hopefully, it will not rain." Many purists have indicated that this is a grammatical abomination. However, the grammarians may be waging a losing war. If the usage becomes firmly established in popular speech, the grammarians may find it expedient to change their rules and call *hopefully* an "absolute" construction, comparable to the Latin ablative absolute. Into the ramifications of this difficulty I do not wish to go at present.

A further problem has to do with metaphors. A *metaphor* implies a comparison (in contrast to the *simile* that makes a comparison explicit). The metaphor conveys an image wherein one term takes on a function originally associated with some other term. For instance, the music critic, reporting on a concert, might say, "The violins tossed the melody to the clarinets," thus comparing the performance on the concert stage to a game of ball. The melody is the ball and the violinists and clarinetists the players. Other common metaphors are: "Her eyes danced mischievously," "He made a brilliant speech," and "His mind immediately grasped the point." *Danced, brilliant,* and *grasped* are used metaphorically. Each offers a comparison between terms that in their literal sense have nothing to do with each other but can relate in a figurative sense.

In these examples the metaphor involves only a single word that conveys a single comparison. In contrast, the comparison may extend over more than one term in a continuous fashion. For a sustained metaphor to be effective, the different terms should harmonize, reinforce each other, and build a coherent composite picture.

A fine example of such an extended metaphor I take from Carlyle. He was speaking of his pleasant childhood and the unhappiness that overtook him as he grew up. First he was quite lyrical about the pleasures of early childhood. Then he described what he felt as he grew older [1].

Green sunny tracts there are still; but intersected by bitter rivulets of tears, here and there stagnating into sour marshes of discontent.

The entire sentence is really a cluster of metaphors held together by a few necessary grammatical terms. All the metaphors harmonize; all contribute to the total picture, with nothing discordant.

Unfortunately, many writers who try to use extended metaphors end with the disastrous product that we call a *mixed metaphor*. Let me give a few examples.

In describing the progress of a musical genius, one author wrote,

His career was really launched in a blaze of critical superlatives after his debut as soloist.

Launching a career is a trite metaphor that in this case compares one stage in a musician's career to one stage in the construction of a ship. But when the author says that a launching took place in a blaze, he has given us a metaphor that is certainly not trite. It is now grotesque.

Another example: The context has to do with a theory that had been under attack. Describing various modifications the theory had undergone, an author declared

Buttressed with this newly acquired varnish of modernity, it. . . .

This is quite unsurpassed as a mixed metaphor. *Buttressed* indicates a massive support that stabilizes a building and calls to mind the great Gothic cathedrals. *Varnish* refers to a type of paint that provides a smooth glossy surface—a superficial covering. Either *buttress* 95 or *varnish* might create an acceptable metaphor. The difficulty lies in the combination: Paint does not make a good support. We may think, perhaps, of the comment, that a ramshackle house was held together by the wall-to-wall carpeting.

One of my favorite mixed metaphors describes the aggressive leader who "took the bit in his teeth and ran with the ball." Another favorite is that of the successful man who "kept his feet on the ground, his head in the clouds, and his nose to the grindstone."

Another stylistic quirk is alliteration. An author driving to get his thoughts down on paper can place together words that repeat a consonant or vowel several times. For example, "The reader will readily recognize the results . . .," "The newspaper provided a platform for his opinions," and "He found himself involved in an intensely interesting experiment." The repetitions of the same consonant (or vowel or syllable) three or more times in successive words I find quite grating.

To be sure, many authors do not seem to mind. Some even try, deliberately, to produce alliterative effects in their prose and resent any criticism of this peculiarity. In my own editing, however, I try to get rid of alliteration whenever I become aware of it. We can usually find a synonym that does not contain the offending letter. Thus, in the first example above we can substitute *easily* for *readily*; in the second, *offered* for *provided*; in the third, *a completely absorbing* for *an intensely interesting*. The difficulty lies not in repairing the fault but in becoming aware of it in the first place.

Only the professional editor (or the teacher) has much opportunity to go over "raw" manuscripts, that is, those that have not been subjected to some sort of correction. Everyone, however, can take his own writings, treat them as the effort of someone else earnestly seeking guidance, and try to effect improvement. You may not be convinced that the text needs change. Your reading may not at first uncover any faults, and you may feel that there is no scope for improvement. However, your reason should tell you that this is not the case, and that if you persevere you can certainly do better.

For those who do not know how to start revising I suggest that you start to rewrite the text and somehow to say it differently. Hunt for synonyms; try alternative constructions; eliminate an adjective here, a preposition there; convert a passive voice to an active; eliminate a whole sentence; and see if any of these changes makes any difference. Weigh the result carefully, and compare the old and the new. As you reflect, perhaps some still different mode of expression will pop into your mind and seem clearly superior. As you gain some critical sense, you will suddenly become aware that your own writing is showing the same faults to which you have become sensitized in the writings of others. The more you ask yourself, How can I say this in a different way? the more quickly will you develop sensitivity and skill in evaluation.

The original writing may have gone fairly rapidly, as thoughts crowded each other in haste to get themselves expressed. The process of revision, however, must go slowly. There is one exception to this. In your first draft, after a couple of hastily written pages of text bring you to a point of rest, you should read over what you have just written, quite rapidly; probably you will find glaring faults that you can correct immediately. The next day you should go over your work much more slowly and much more critically. You no longer will be under the pressure of getting down on paper what you wanted to say, and you can devote much more attention to the way you are saying it. You can then, without interrupting your train of thought, keep asking yourself whether you can say it better. You may decide that what you have written needs no change at all, but you should not make such a judgment until you have at least tried some alternatives. I would bring to mind the example I gave earlier, of trying to find the best arrangement of furniture. You should at least think of alternatives, even if you do not adopt them.

Revision, I have said, is the process of editing your own work. There is, however, an important point of difference. If, when engaged in editing, you feel that major changes are in order, you cannot be sure that any alterations you propose will express what the author wanted to say. You may be distorting his meaning. In revision, however, you are in control at all times. You have complete freedom to make all the changes you want.

The problems of revision fall into two distinct categories, separable   97 yet related. One I would call *style* in the narrow sense, having to do with the way things are said—the choice of words, the turn of a phrase, the grammatical constructions—the various aspects of writing that can be illustrated by short quotations. Many of these features I have discussed in earlier chapters and in this chapter my comments all relate to this component. When we apply these various recommendations to our own writing, we have revision, but only in a restricted sense.

In contrast, we have the problems of total *organization*, which involve the composition as a whole, whether it be a 500-word essay or a 150,000-word book. Any composition, whether short or long, must have a unified structure. A single essay must have a unity. A book as a whole must have a unity. A book will have many chapters, each of which in turn must exhibit unity. An author must decide not only what he wants to say (as well as how to say it) but also where to put it. What should he leave out? What emphasis should he give this or that portion? What sort of balance will result, and how will one or another part affect the clarity of the whole? These are some of the factors that revision must take into account.

So far, in discussing revision, I have emphasized the stylistic aspects. The organizational problems I will take up in an appropriate place in the next chapter.

References

1. Carlyle, T. *Sartor Resartus.* London: Dent (Everyman's Library), 1973. P. 78.
2. Rublowsky, J. *Pop Art.* New York: Basic Books, 1965. P. 131.

Getting Finished

If you have engaged in serious writing, you probably had no trouble in getting started but great difficulty in bringing the work to completion. Undoubtedly you began with enthusiasm, yet this soon waned, the work dragged, became more and more of an effort, and finally was put aside to be finished "later." But perhaps "later" never came.

Whatever you want to write—a brief report for a local journal, something more ambitious for a major periodical, or perhaps a monograph or a book—requires preparation. Let us assume that the preliminary work has been done, notes accumulated, a rough outline prepared, and at least part committed to paper. The problem is, to finish it all.

The word *finish* implies a three-fold division, with a beginning, a middle part, and then the completion—the real end. The middle portion is the area of writing and rewriting, of sweating and agonizing. If you kept this up long enough, you might decide that you at last had come to a point where you could call the work finished. However, I can suggest two kinds of *finish*, or perhaps two and one-half. The finish in its first sense arrives when you decide the manuscript is ready to send off to an editor or publisher. But the editor, unless he rejects it outright, will usually suggest many changes that he thinks will improve it. After these have been made and the paper accepted, the manuscript is subjected to copyediting, which introduces further changes. Finally, after much tribulation, it appears in print. Only at that point is the paper finished in the second sense.

However, when you have read over what appeared in print, you may think to yourself how much better the work would have been if only you had expressed yourself in *this* way instead of *that*. Such *esprit de l'escalier* may provide inward gratification but will have no effect on the objective product. Hence, I allow it only a half-stage in the process of finishing.

For practical purposes we must concentrate on the first stage—getting the manuscript into such shape that you are willing to send it off to the editor or publisher. Let us study some of the pitfalls that may entrap you before you reach your goal.

When Is the Manuscript Finished?

Revision is the operative process that leads to completion. After you get your ideas on paper, you face the two questions primary in all literary effort: Am I really saying what I want to say? and Can I say it better? As someone has declared, there is no such thing as good writing, only good rewriting.

Essential for successful writing is the awareness that I would call artistic conscience. The writer's artistic sensitivity tells him, "No, that is not good enough, and I must make it better." This whispering of conscience may apply to the ideas he is trying to convey or to the way that the ideas are actually expressed. I thus distinguish the larger configurations of thought from the smaller units of expression—the conceptual content from the phrases and sentences that convey it. The first aspect represents the message, the second refers to style. The relation between the two is, of course, close, for thoughts must be expressed in a suitable style, and sentences must have an adequate content.

In revision, the sensitive author will try to simplify the text, make his language more precise, identify gaps in his exposition and recognize any concepts that simply do not "fit" the rest of the work. He will be attending to both organization and style. In theory, when these can no longer be improved, he has completed his revision and his writing is finished.

However, such a theory is entirely quixotic. It aims at perfection, and perfection is an asymptote, a limit not to be reached in finite time. Let us examine in realistic fashion some of the factors that hinder revision and prevent the completion of a manuscript.

The Search for Perfection

One hindrance is a drive for perfection that may attain an obsessive status. A splendid example of this we find in Camus' *The Plague*. One of the characters, M. Grand, was writing a book, but he could not get beyond the first paragraph. He wanted the work to be flawless and he rewrote the paragraph endlessly, spending perhaps a whole day meditating whether a given word expressed exactly the sense that he had in mind. Never quite satisfied, he kept changing a word or a phrase. And he would not proceed to the second paragraph until the first was perfect. His manuscript had attained a length of about 50 pages, but it consisted principally of the same first sentence, written over and over with small variations and elaborations.

The Plague is highly symbolic. Nevertheless, however figurative the incident, M. Grand does manifest a trait that is all too common: a need for perfection that will forever prevent a work from getting finished. We can think of this as a personality disorder rather than a stylistic merit.

Other types of personality disorder may also prevent an author from bringing a work to a conclusion. The great American historian F.J. Turner built up a variety of excuses for not completing the work on which he was engaged. He would rewrite drafts endlessly; he would delay while looking for one extra bit of evidence, for one further fact; he would flit from one subject to another, using the momentary pleasure of further research as a way of avoiding the discipline of writing. He would indulge in "overresearch and overpreparation," and although he had accumulated vast masses of notes, there were always good reasons why he was unable to finish a particular work. At the same time he would be signing additional contracts for books that he never wrote [1].

Encountering and Overcoming a Blockade

Although few people have this trait to the extreme degree that Turner did, many writers have trouble in bringing their work to a satisfactory conclusion. We may think of this as one form of writing block.

Usually a block appears not at the beginning of a work but only after a considerable portion has already been written. The author suddenly finds that he has nothing more to say. Everything has dried up. The flow of words and ideas stops. Of course, if the author really has nothing to say, he should not be writing in the first place, but we presume that there is a lot in his mind that simply won't emerge. The associations seem to have been cut off, and he does not know what to do next.

I suggest two types of block. One I call the stone wall and the other the floundering around. In the first the cessation of ideas is, so to speak, absolute, and the author cannot go anyplace. In the second type the block is relative: Forward progress is obstructed, but the author goes sideways, saying the same thing in different words when he wants to say something new. These two types require rather different forms of treatment.

Sometimes the stone wall block may have deep psychological bases, but for the most part it does not. Although I am not able to identify any underlying causes, I can make suggestions for breaking the blockade. You cannot batter down the wall by sheer force, but you can try to get around it by going off in a totally different direction; or you can try to chip away a little of the mortar and make a small crack that will eventually enlarge and permit you to demolish the wall with relative ease.

To go around the wall, you leave whatever topic you are working on and make a fresh start on something quite different. If, for example, you are writing a book and get blocked part way through, you can start on a new chapter, ignoring the blockaded path. After traveling along the new path for a while and establishing a nice flow of ideas in a different direction, you can go back to the blocked portion. You will probably find that most of the wall will have tumbled down by itself and that the remainder causes you little difficulty. Sometimes, when the going is rather tough, you may find yourself working on two or three chapters at a time, passing from one to another according to the flow of ideas. At any sign of impedance you can shift to another chapter without waiting for the flow to come to a total stop.

Sometimes, however, there is no other place to go. You may be working on the last chapter of the book or else writing a paper that cannot be put aside. You must try to get through that wall, somehow. There is a biphasic technique—first, reflect, and second, write. The reflection may not at first seem productive. You take your notes and start to go over them again—and again—and again. Perhaps some of the thoughts you had assigned to one section of your work will appear discordant, and you will want to transfer them to another portion; or rereading some notes may bring to mind a small point that can set off a whole new train of thought. Perhaps you will suddenly realize that you do not have enough evidence on one particular point and you will want to go back to your sources for confirmation or amplification. If nothing like this occurs, do not be discouraged, but keep on meditating. Start again the next day. When thoughts come, as sooner or later they will, jot them down immediately. Do not try to exploit them. For the moment, be content to get them safely on paper.

The meditation you can alternate with writing. Even if you do most of your writing on the typewriter, I recommend scribbling by hand when you are trying to overcome a block. Start with a relevant note, or a thought that you had jotted down, or a couple of sentences from the last page or two of the manuscript. And keep the pen moving. You can start by copying what you have already written, writing it again and again, perhaps with an addition of another sentence or two. As you copy, try to make some change in the wording, no matter how small. Try a variation. And the next time you copy it, make a bigger variation—adding a couple of words, finding synonyms, changing the grammatical construction—what you do is not important, so long as you make some sort of change. And soon there will appear a new thought, not necessarily along the same lines you had been working on, but relevant thereto.

A chain of association has been reestablished, although on a tenuous basis. You have achieved a new association and not merely a rewording of the old. You must nurture carefully the new thought, and let it lead you wherever it wants to go. But keep on writing, without worrying about elegance of expression or even grammar. *Keep the pen moving.* You must not worry if the new associations do not have much to do with the original theme from which they arose. They will serve to prime the pump and release a vigorous flow. If that flow leads you away from the original direction, let it carry you along, and after a while you will find that you have a new segment of writing. If it happens to follow the same direction as the original, you are unusually lucky, for most often the break in the blockade will give you something at a distinct tangent, whose "fit" with the remainder may pose a further problem. This new problem, however, will be easier to deal with.

At this point you are encountering the second type of blockade, which I have called floundering around. You have got rid of one difficulty only to encounter a second, which actually is far more common than the absolute block. Floundering around represents a state of confusion in the author's mind. His thoughts are not clear to himself and hence will not be clear to any reader. Some authors, whose writings are entirely opaque, write and then publish what they write, without any apparent concern with its intelligibility. Such authors we will ignore, for they are not even aware that they are floundering. What I call floundering takes place when an author has a certain degree of artistic conscience and realizes that what he is putting on paper is not good enough.

As a simple example we can take the writer who, after expressing a thought, apparently has some degree of inner discomfort. He then starts a new sentence by saying, "In other words . . ." and proceeds to give the same thought in different form. Not infrequently he starts still another sentence by "Or, differently stated . . ." and again gives the same idea in still different words.

The author is aware that he has not expressed himself adequately, and, by repeating his exposition in different words, he is attempting to clarify his ideas in his own mind. Only during the process of writing does the author achieve a full understanding of his own thought. The phrase "In other words" is a good sign that he realized he was floundering and tried to achieve clarity. His only mistake was in letting the floundering appear in print. After he had reached some sort of resolution through reworking his ideas, he should then have discarded all he had written and replaced it by a simpler and more compact expression, unequivocally clear (and probably about a third as long). The phrase "in other words" appearing in a text tells us that the author has some artistic conscience but not quite enough.

A somewhat different example we find in prose that is written rapidly. Sometimes an author is in a creative frenzy. If his mind is working at high speed, chains of association may present themselves rapidly, pressuring each other to get on paper. As he writes, the connections between thoughts seem beautifully clear. Yet a few days later, when the creative frenzy has passed and he reads over the text, he sees numerous gaps in the reasoning, while transitions seem faulty and thoughts disconnected. The text does not possess a unity. The ideas that seemed blindingly clear at one time now appear murky.

Writing that flows easily may, then, lack consistency. The conscientious author will recognize this, carefully examine his text, and discard those ideas that do not harmonize with the main flow. He should become aware of discordance—a lack of what I call "fit"—and perceive faults in organization. When the organization lacks harmony, the conscientious author will find himself unhappily floundering; on the other hand, the author who lacks an artistic conscience will not care, and his writing will be correspondingly dreadful.

Sometimes, as an author pursues a train of thought, he finds himself in a blind alley. Realizing that he is temporarily blocked, he may put aside the pages that are causing the difficulty, start on a different train of thought, and continue for several more pages. Then he has the task of combining the different installments of his text so that they will read smoothly. Sometimes he cannot get the segments together, cannot find the connections with which to link together the different parts. Convinced that the associations exist but lie buried in his subconscious, he starts to dredge. His efforts may have either of two outcomes. He may discover the hidden connection, pinpoint the gap in his exposition, and realize what he must do to fill it in. In this way he will have reestablished a smooth progression from one part to another and restored the unity of his text.

On the other hand, he may continue to flounder until he gets a sudden illumination: the part that seems discordant *is* discordant; the only cure is to eliminate it. He discards a portion as not relevant to what he is trying to say. And at once the smooth flow of ideas is reestablished, the unity restored.

Discarding portions of a manuscript is an act of renunciation that may cause the inexperienced writer great anguish. But it is a sacrifice on a worthwhile altar—the altar of organic unity. The discarded portion need not be thrown away. If placed in a "miscellaneous" file it may serve eventually as a nucleus around which later associations may cluster, and perhaps eventually lead to some new creative thought.

Trying to say too much is a common fault. When an author has done a great deal of reading and thinking, he may feel that all his work must get reflected in the finished product. Such a compulsion, such a dread of leaving something out, can be a major cause of floundering. If an author actually uses 30 to 40 percent of the notes he has accumulated, he has done well. The remainder has enriched his background and furnished perspective. It is not wasted.

Sometimes, when the floundering is quite severe and the author cannot find any way to emerge from the confusion, a more drastic method may help. To any writer who gets mired down, unable to proceed, I recommend this technique: Go over what you have written and summarize each paragraph in a single sentence. If you have trouble in so doing, then probably your paragraphs are poorly constructed and contain discordant ideas. However, do not stop to revise. If necessary, put down two sentences for a paragraph, with the mental note that here is a spot that needs fixing, but do not try to fix it now.

Carry this procedure through the entire portion of the text that is giving trouble. You will then have a series of topic sentences that indicate, in a stripped-down fashion, what you actually have said. This will probably differ markedly from what you think you said. The rhetoric that elaborates a thought may have obscured the thought itself. This procedure, of making topic sentences, separates the thoughts from the rhetoric (and when you are done you may be surprised at the paucity of thought).

When you read over the series of topic sentences, you can easily see if there is a logical continuity of ideas or a faulty development; you can spot digressions and appreciate weaknesses; you can see where further exposition is required. You can readily distinguish the major and the minor subdivisions. You can then regroup the topic sentences, delete some altogether, and transfer others to a different locus.

By reducing your work to topic sentences you can analyze the organization and the changes you need to render it more suitable. When the rearrangement finally seems appropriate, then you must rewrite. You may have difficulty, but at least you know where you are trying to go.

When a writer finds that he is going round and round without making any progress, he may try another mode of getting out of the morass. By talking things over with a friendly critic (such as a sympathetic spouse), he can try to express orally what he was laboriously trying to put on paper. If he has devoted enough thought to the subject, and is lucky, there may suddenly emerge from his subconscious a simple clear statement of his ideas. And a great light may dawn—"*That* was what I was trying to say all the time; why could I not have said it before?" The answer: The concepts had not been sufficiently digested. The act of writing long pages of unclear prose had been part of the process of digestion. Clearer understanding suddenly came as he made the simplest possible oral presentation. At this point he should quickly jot down those few sentences, and all the rest will be easy. He discards the obscure pages and rewrites it all much more simply and compactly. And the floundering has given way to a smooth assurance.

I can summarize the requirements for good expository writing in a single sentence: Know what you want to say, say it clearly, and then stop. But such an exhortation is extremely difficult to carry out. I find that I am never quite sure what I really want to say until I am in the process of saying it. I often get bogged down and must then extricate myself by clarifying my thoughts according to some of the precepts discussed here.

Shortening the Manuscript

Knowing when to stop sounds easy but in practice may be difficult. The inability to stop is part of the common fault of *wordiness*. Let me comment briefly on this fault. We can approach the topic by thinking of conversation. We all recognize the dreadful bore who talks in endless detail and does not distinguish the important from the trivial. He gives the most minute circumstances, even if quite irrelevant. A little different, but also a bore, is the speaker who keeps going on and on, saying the same thing over and over again in slightly different words, and not knowing where or when to stop. Writers who suffer from these faults may make the excuse that time was pressing, an excuse that brings to mind the closing sentence of a letter—attributed to various authors—"If I had more time, I would write you a shorter letter." To ramble at considerable length is easy; to provide a disciplined order is hard, requiring thought, effort, and skill.

As an editor I have often received manuscripts that deserved to be published but were far too long. A suitable descriptive term 109 would be *windy*. Often I have suggested to the author that his paper would be acceptable if reduced, say, from 15 pages to 12. Occasionally I have received the "revised" version completely retyped and filling 12 pages. The text remained virtually unchanged, however, but the *margins* were vastly smaller! By this maneuver the author had literally carried out my suggestion and returned to me 12 pages of text instead of the original 15.

Another form of shortening is massive deletion, removing whole paragraphs while leaving the intervening text without change. The deletions, in the aggregate enough to bring the pages down to the requested total, nevertheless produce only mininal improvement. If the overall style is flabby and edematous, then in spite of the deletion the residual text remains flabby and edematous. "Revision" of this type calls to mind the different possible ways of losing weight. If a physician tells a patient that he must lose 40 pounds, one way of carrying out the doctor's orders is to amputate a leg. This, however, while it eliminates a certain number of pounds, does not really improve the health of the patient. The preferable alternative is to go on a diet.

Dieting is hard work and not very pleasant, and so too is "slimming" a manuscript. It requires careful attention to detail, a line-by-line and word-by-word scrutiny. Early in my professional career I had a most salutary experience that impressed this point upon me. My first major research project, involving two years of work, I had presented in a manuscript of about 40 pages. This I sent to the one journal that specialized in papers of this subject matter. The editor, after a decent interval, wrote back that he would gladly publish the manuscript if I would cut it precisely in half.

The letter was indeed discouraging, for I had worked hard on the writing and thought it quite readable. To be sure, as I reread it I admitted that if scattered sentences were to disappear, no harm would result. But this simple procedure would not achieve the needed shortening. Something more drastic was necessary. Over a period of some two weeks I subjected the paper to a line-by-line critique, eliminating single words, finding a short word to substitute for a long one, a single word that might take the place of two, a short phrase to replace an entire clause, or a single sentence to express the gist of an entire paragraph.

The work of revision was exhausting and time-consuming, but I finally did get the text down to half its original length. And I was amazed to see that the result was incomparably better. Then for the first time did I truly appreciate the force of the epigram, "Half as long is twice as good."

As an editor I speedily learned that merely telling an author to shorten his paper rarely had the desired effect. Indeed, verbal exhortation of any sort usually produced little result. Example is better then precept. If I wanted an author to shorten his paper, I would usually take a page or two (in a photo-duplicated copy) and subject it to the line-by-line revision that I wanted him to make, demonstrating how a word or phrase or clause could vanish, with actual improvement of the remainder. And I would carry through other editorial changes of the type described in the preceding chapters. I would then urge the author to continue the process with the rest of the paper. Sometimes he would do so. Sometimes I never heard from him again.

In closing this chapter I would offer one word of advice to authors: Before you consider your work finished, go over it carefully to see how much you can shorten it—find short words to replace long ones, delete redundant phrases and clauses, cut out whole sentences if they are merely rhetorical ornament, make your writing simple. In one writing seminar a young physician remarked, after I gave comparable advice, "If I made my paper as simple as you want me to, no editor would publish it." What reply would you make to this comment?

Reference

1. Hofstadter, R. *The Progressive Historians: Turner, Beard, Parrington.* New York: Vintage Books, 1970. Pp. 115–116.

7

Style Analysis

Defining Style

Ordinarily we have little difficulty in recognizing friends and relatives and, even at some little distance, distinguishing them from strangers. If asked, How do you identify a friend? What are the mental processes involved? we would be rather hard pressed to find an answer. We would probably say, in effect, that we have an "intuitive" reaction; and if we try to go beyond this vague reply and identify the perceptual cues involved, we might point to bodily configuration, walk, mode of dress, and the like. Making the discrimination might be easy, but giving the reasons on which we made the judgment is not.

The situation is similar with other modes of discrimination. Suppose we listened to two different pieces of music, neither of which we had ever heard before, one written in the seventeenth century, the other in the late twentieth. If asked, Which is the modern piece? we would have no difficulty in answering. Likewise, if confronted with a Renaissance painting and a modern painting, or an Elizabethan poem and some modern verse, we would have no trouble in discriminating between them, even though they were both completely new to us. We have enough general familiarity with artistic modes that when we must discriminate between something new and something old, we can usually say, *That* has a modern ring to it. And yet we would find difficulty putting into words the reason for our decisions.

If the test involves deciding between two examples of widely different characteristics, the task is quite easy. But think for a moment of the art expert, given a single painting that he has never seen before, who must decide the approximate date when it was painted and the identity of the artist. The expert would know the characteristics that distinguish one era or school from another and, within a given era or school, one artist from another. Ultimately he would probably rely on intuition; after making careful studies he would look at the picture, and look some more, and finally say, It resembles the work of So-and-so. But although the expert might ultimately rely on intuition, he would back this up with specific reasons. He would make explicit the grounds on which he attributed the work to painter A rather than painter B, so that other experts could have a basis for agreement or disagreement.

In all these problems, from the simplest to the most complex, the 113 judgment rests on a knowledge of *style*, a knowledge highly detailed and explicit in the expert, vague and diffuse in most of us. We can apply the same concept to the printed word. Take, for example, the essays called *The Federalist*, published in 1787–1788, so influential in bringing about the ratification of the Constitution. We know that they were all written by Alexander Hamilton, James Madison, and John Jay. We know the specific authorship for most but not all of the essays. In those works whose author is not definitely known literary critics have tried to supply an answer by examining internal evidence, that is, studying the style of the essays in question and comparing the different features with those that characterize each of the three writers. The experts cannot rely merely on dumb intuition; they must give reasons and evidence; and to do this they must be able to analyze the prose in explicit fashion.

What is the *style* that they are trying to analyze? I would offer a definition: Style is the aggregate of qualities that, relative to some particular activity, allows us to discriminate between one person (or group of persons) and some other person (or group). We can distinguish the style of Dürer from that of Bosch, of Picasso from that of the Barbizon school. We can speak of the style of the Imagist poets and contrast it with that of the metaphysical poets; and distinguish the writing of Jonathan Swift from that of Henry James. To make judgments we rely on certain qualities that provide the (approximate) specificity.

Whoever wants to improve his own writing can have no better exercise than to analyze the style of whatever he reads. He should try to identify the qualities that characterize the different kinds of writing with which he comes in contact. For example, why are the directions on the income tax forms so difficult to follow? In the daily newspaper, how does the writing on the sports page differ from that of the editorial page? How does a case report in a medical journal differ stylistically from a short story in a popular magazine? That they do differ is obvious, but it is not so easy to identify the particular qualities that account for the difference.

Writers on writing often give this advice: read the great classics of literature, the "good" authors, and you will automatically absorb the ability to write well yourself. With all due respect for the eminent writers who have offered this advice, I consider it essentially nonsense, remarkably unhelpful to the struggling author who tries to improve his own communicative skills. What the aspiring author needs is a critical sense and the ability to make discriminations. These he acquires only by deliberate effort, not by passive osmosis.

The aspirant must constantly examine whatever he reads, whether a newspaper, an advertisement, a theater program, a medical journal, or current works of fiction or nonfiction that he reads for pleasure or instruction. And he must constantly ask himself, Do I like it? This query usually has within it a series of overtones—Is it easy to understand? Is it effective? Is it worth saying? Does it grab my attention? Then the aspirant must try to answer the second major question, What accounts for this reaction on my part?

Constant attention to these two queries leads us to the heart of style, for by our answers we can identify qualities of writing. And once we have skill in identifying individual qualities, we can aggregate them into bundles that characterize one or another *kind* of writing. And with further practice we learn to distinguish individual authors according to the combination of qualities that their writings exhibit.

In this chapter I give some examples of strikingly different styles, relatively easy to characterize and differing markedly one from the other.

Seventeenth Century Examples

Even in the seventeenth century critics declared that the quality of scientific writing needed improvement. Of course, at that time, there were no professional scientists in the sense that we understand today. In the latter seventeenth century most investigators were amateurs for whom the term *virtuoso* was the popular designation. In 1660 the founding of the Royal Society marked an important step in the organization of scientific activity, and in 1666 the initiation of their journal, *The Philosophical Transactions of the Royal Society*, marked the beginning of modern scientific periodical literature. And the style of writing began to show substantial changes from the literary mode.

Earlier in the century English prose had a rather characteristic wordiness and a convoluted syntax. We think immediately of John   115

Milton and Sir Thomas Browne. Let me give two brief examples from Browne's writing [2]. In the first passage he discusses the credulity of man as a cause of error. By credulity he means the

> *believing at first ear, what is delivered by others. This is a weakness in the understanding, without examination assenting unto things, which from the Natures and Causes do carry no perswasion; whereby men often swallow falsities for truths, dubiosities for certainties, feasibilities for possibilities, and things impossible as possibilities themselves. Which, through the weakness of the Intellect, and most discoverable in vulgar heads; yet hath it sometimes fallen upon wiser brains, and great advancers of Truth.*

In a second passage he discusses the belief that the legs of elephants have no joints.

> *The hint and ground of this opinion might be the gross and somewhat Cylindrical composure of the legs, the equality and less perceptible disposure of the joints, especially in the former legs of this Animal; they appearing when he standeth, like Pillars of flesh, without any evidence of articulation.*

We must realize that many of the words are used in a sense different from that of today. Thus, *dubiosity* means *doubt, former* means *front, composure* indicates *composition* or *structure*, and so on. But aside from obsolete usage, the style shows overelaboration and redundancy. Browne generally wrote in complicated sentences with many abstract polysyllabic words. He liked to play with an idea, say it over again in different ways, worry it a little in playful fashion. He was a physician, a man of wit and high intellectual attainments, but he was not a virtuoso, not a scientist.

If we keep in mind this type of writing, we can appreciate Robert Boyle's views of the style appropriate for scientific communications. In one of his early essays, published in 1661 [1], he declared,

. . . where our design is only to inform readers, not to delight or persuade

them, perspicuity ought to be esteemed at least one of the best qualifications of a style; and to affect needless rhetorical ornaments in setting down an experiment, or explicating something abstruse in nature, were little less improper, than it were (for him that designs not to look directly upon the sun itself) to paint the eyeglasses of a telescope, whose clearness is their commendation, and in which even the most delightful colours cannot so much please the eye, as they would hinder the sight.

The comparison between verbal elaborations and the decorating of the lens of a telescope emphasizes his point, that in scientific description clarity is the greatest virtue and that painting unnecessary word-pictures and affecting "needless rhetorical ornaments" are drawbacks. He is condemning useless figures of speech and verbal elaborations, and praising simplicity and clarity.

In studying this passage we are struck by the rhetorical ornaments that he supposedly rejects and the masses of words that he uses to express his thought. He is saying, in effect, that if we want to convey information to readers, we should write simply and clearly—the same advice that is so abundantly offered in the late twentieth century. But advice of this character had as little effect in the seventeenth century as it does in the twentieth.

In 1667 Thomas Sprat wrote a *History of the Royal Society of London*, in which he not only described the history of the society but also indicated some of the benefits produced thereby. The Society wanted to correct the "excesses in natural philosophy," one of which was an elaborate, highly convoluted manner of writing. To remedy the situation the Society was [14]

most rigorous in putting in execution, the only remedy, that can be found for this extravagance: and that had been, a constant resolution, to reject all the amplifications, digressions, and swellings of style: to return back to the primitive purity, and shortness, when men deliver'd so many things, *almost in an equal number of* words. *They have extracted from all their members, a close, naked, natural way of speaking; positive expressions; clear senses; a native easiness: bringing all things as near the mathematical plainness, as they can: and preferring the language of artizans, countrymen, and merchants, before that, of wits, or scholars.*
[Original capitalization omitted; spelling and punctuation retained.] 117

But alas for good advice! "Amplifications, digressions, and swellings" are themselves an amplification, digression, and swelling. The whole passage is long-winded and repetitive.

Eventually the journal of the Royal Society, *The Philosophical Transactions*, did have a considerable effect on scientific publications in English, but the exhortations of the Society were certainly not immediately effective. Joseph Glanvill, a clergyman member of the Royal Society, published a book in 1668, in which he referred to Sprat's *History*. Said Glanvill [5],

> *For their History, that is newly come abroad, gives so full and so accurate an account of them and their designs, that perhaps it may be superfluous to do more in this, than to recommend that excellent discourse to your perusal, which I do with some more than ordinary zeal and concernment, both because the subject is one of the most weighty and considerable that ever afforded matter to a philosophical pen, and because it is writ in a way of so judicious a gravity, and so prudent and modest an expression, with so much clearness of sense, and such a natural fluency of genuine eloquence: so that I know it will both profit and entertain you.* [I have omitted the italics and the capricious capitalizations of the original.]

Glanvill, despite his lip service to the precepts of the Royal Society, wrote on the principle, Why say things simply if you can just as easily say them in a wordy and complex manner? There are too many words, most of which contribute little and detract a lot. Instead of "clearness of sense and fluency of genuine eloquence" he might have said "clearly and fluently." Instead of "recommend that excellent discourse to your perusal" he might have said "recommend it."

But here we encounter an important point: Manners of speech are deeply ingrained and cannot be changed merely by good advice. The long-winded modes of expression were deeply rooted in the culture of the seventeenth century. In time a simpler and more graceful style, under the leadership of authors like Dryden, Addison, Swift, and Steele, struggled for dominance, but the change had to 118 come slowly. For scientific writing to improve, a great deal of effort was necessary, and most of the writers did not make any specific effort.

Macaulay and Carlyle

The history of style is an engrossing subject, far too large to take up here. After these few examples of seventeenth-century writing offered here with only brief comment, I will jump to the mid-nineteenth century and take up two contrasting literary figures, Macaulay and Carlyle, whose writings lend themselves readily to specific stylistic analysis. Even though neither science nor medicine is involved, the dissection can help us with our present-day stylistic problems.

Some writers put their words together in such a way that a reader, who has never seen a particular passage before, but has some general knowledge of English literature, can make a fairly good guess as to the author's identity. Macaulay is such an author. Yet when we try to analyze his style we must not assume that his writing is uniform throughout. Any writer may show considerable variation from one work to another, from one period in his life to another, and from one type of prose to another.

Let us look at a few passages from different works of Macaulay. In this first example he is discussing Oliver Goldsmith [8].

Minds differ as rivers differ. There are transparent and sparkling rivers from which it is delightful to drink as they flow; to such rivers the minds of such men as Burke and Johnson may be compared. But there are rivers of which the water when first drawn is turbid and noisome, but becomes pellucid as crystal, and delicious to the taste, if it be suffered to stand till it has deposited a sediment; and such a river is a type of the mind of Goldsmith. His first thoughts on every subject were confused even to absurdity; but they required only a little time to work themselves clear. . . .

This is effective writing. It conveys an image, a vigorous picture, although whether that picture is a true one, we must leave to historians. Here we are concerned only with the way Macaulay has put 119 his words together to achieve his rather distinctive style.

We note first of all an extensive use of simile and metaphor. He makes an explicit comparison between minds and rivers and then carries through the comparison in several different respects—the clarity of the water, its taste, the sediment that it might deposit, the process of self-cleaning. In Chapter 5 we saw some of the absurdities that can result when an unskillful writer tries to carry through an extended figure of speech. In Macaulay's passage, however, the metaphors hold. Each facet serves to emphasize the comparison. There is no mixing of the metaphor, and each part of the extended figure is apt.

Then there is a certain hypnotic rhythm. The sentences appear to vary markedly in length, but they are either broken by semicolons or are compound sentences with a coordinate conjunction. Actually, they have the effect of a series of relatively short simple sentences, lightly joined. We readily perceive this if we read the passage aloud, dropping the voice slightly at each semicolon and every time there is an *and* or a *but* preceded by a comma. When we do this we find a succession of simple sentences or quasi-sentences, with an occasional complex sentence thrown in.

This can lead to a certain monotony, both of structure and of rhythm. He achieves a balance through repetition and antithesis—a sort of continuing "on the one hand . . . but on the other." The water is turbid, but it becomes clear. Goldsmith's thoughts were confused, but they too become clear. When this antithetical balance continues for paragraph after paragraph, the net result can be quite tiresome, however effective it may be in small doses. Sometimes when we read Macaulay we get the feeling of a full-rigged sailing ship, rolling from side to side. When we roll to one side, we expect to be carried back . . . and roll again . . . and back . . . and roll again. I would compare Macaulay to a galleon under full sail. The symmetry, sometimes obvious but sometimes subtle and concealed, gives a certain stately quality to the prose, but a quality that can become monotonous.

In this passage I would emphasize the use of adjectives—*transparent, sparkling, delightful, turbid, noisome, pellucid, delicious,* and so on. The color and force of the writing comes from this masterful use of adjectives, while the verbs are relatively drab, with little force. Yet Macaulay has used his adjectives skillfully and avoids the sense of overloading. We do not have that deadly property whereby each noun is encumbered with one or two adjectives. Instead, the adjectives are in large part complements rather than direct modifiers, or else they follow the noun.

Let us examine another passage of a different quality. This is historical narrative, describing the flight of Mary, the wife of James II, together with their infant son [9].

> *The party stole down the back stairs, and embarked in an open skiff. It was a miserable voyage. The night was bleak: the rain fell: the wind roared: the water was rough: at length the boat reached Lambeth; and the fugitives landed near an inn, where a coach and horses were in waiting. Some time elapsed before the horses could be harnessed. Mary, afraid that her face might be known, would not enter the house. She remained with her child, cowering for shelter from the storm under the tower of Lambeth Church, and distracted by terror whenever the ostler approached her with his lantern.*

Here we have effective narration, couched in staccato sentences coming in rapid succession like bursts from a machine gun. We do not have the antitheses of the previous passage, the stately rolling back and forth, but we do have a definite rhythm, quite effective in building up suspense.

Then, too, the choice of words is excellent. Macaulay has used forceful adjectives—*miserable, bleak, rough*; and also forceful verbs—*stole, embarked, roared.* Strong color alternates with more prosaic words that fill in the picture.

With this brief example in mind, let us go back to the expository style. I now give an example that compares the two great British political figures, Fox and Pitt [7].

> *The speeches of Fox owe a great part of their charm to that warmth and softness of heart, that sympathy with human suffering, that admiration for everything great and beautiful, and that hatred of cruelty and injustice, which interest and delight us even in the most defective reports. No person, on the other hand, could hear Pitt without perceiving him to be a man of high, intrepid, and commanding spirit, proudly conscious of his own rectitude and of his own intellectual superiority, incapable of the low vices of fear and envy, but too prone to feel and to show disdain.* **121**

Here we see a return to the rolling antithesis, and we even have an express *on the other hand*. The sentences are rather long, yet their length is due not to a succession of clauses but to a piling up of adjectives and modifying phrases. The whole quotation has only two sentences. The first is complex, with the main verb *owe* and a single subordinate clause with a compound verb, *interest and delight*. The second sentence is simple, with a single subject, *person*, and a single verb, *could hear*. The effect of the passage derives from the succession of nouns and adjectives, used separately or in combination. Sometimes the nouns have preceding adjectives—*high, intrepid, and commanding spirit* and *intellectual superiority*—but the combinations do not become monotonous. However, we must note the tendency of "doubling"—using two different terms with but minor differences between them: *great and beautiful, cruelty and injustice, interest and delight*. This can indeed get monotonous.

In a last example Macaulay is describing the young Samuel Johnson and contrasting his appearance and his mind [10].

> *His cheeks were deeply scarred. He lost for a time the sight of one eye; and he saw but very imperfectly with the other. But the force of his mind overcame every impediment. Indolent as he was, he acquired knowledge with such ease and rapidity, that at every school to which he was sent he was soon the best scholar.*

Here, too, although in a somewhat more subtle fashion, we see the balance and the antithesis. There is the same tendency to short simple sentences, with the occasional interposition of complex sentences of greater length. Even the last sentence, with a main clause and two dependent clauses, conveys an antithesis—he was indolent but he readily acquired knowledge. The vigor of this passage does not derive from any one part of speech but inheres in nouns, modifiers, and verbs, in fair balance.

In my writing seminars I always gave the students an excerpt or two from Carlyle and asked them the simple questions, Is it good writing or bad? Do you like it or not? Invariably the majority thought the writing was bad, even though an occasional student might express appreciation. After discussion and careful analysis, however, most of the class would recognize the merits and alter their judgments.

Today very few read Carlyle for pleasure; ordinarily he is read only under the spur of a college assignment. And since, over the years, not one single member of my classes has ever identified Carlyle's unique style, we may guess that very few physicians have been exposed to him.

A liking for Carlyle is definitely an acquired taste. Most modern readers, at first exposure, find his style rather unpleasant, even repulsive. Yet once we get used to his idiosyncrasies, and allow for the unevenness of his writings, we must recognize him as one of the great masters of English prose. Students who seek earnestly to improve their own writing style would do well to study Carlyle attentively.

I will start with a quotation from *The French Revolution*, in which Carlyle discussed the approaching bankruptcy of the kingdom [3].

How singular this perpetual distress of the royal treasury! And yet it is a thing not more incredible than undeniable. A thing mournfully true; the stumbling-block on which all Ministers successively stumble, and fall. Be it "want of fiscal genius," or some far other want, there is the palpablest discrepancy between Revenue and Expenditure; a Deficit *of the Revenue: you must "choke (*combler*) the Deficit," or else it will swallow you! This is the stern problem; hopeless seemingly as squaring of the circle. . . . Are we breaking down, then, into the black horrors of NATIONAL BANKRUPTCY?*

Great is Bankruptcy: the great bottomless gulf into which all Falsehoods, public and private, do sink, disappearing; . . . For Nature is true and not a lie. No lie you can speak or act but it will come, after longer or shorter circulation, like a Bill drawn on Nature's Reality, and be presented there for payment,—with the answer, No effects. *Pity only that it often had so long a circulation: that the original forger were so seldom he who bore the final smart of it. Lies, and the burden of evil they bring, are passed on; shifted from back to back, and from rank to rank; and so land ultimately on the dumb lowest rank, who with spade and mattock, with sore heart and empty wallet, daily come into contact with reality, and can pass the cheat no further.*

Although in the quotations from the seventeenth century I deleted the idiosyncratic capitalizations and italics, I have preserved them   123 in the passage from Carlyle. These diverge from the usual nineteenth-century standards and form part of Carlyle's special style.

As we read the quotation, several things come to mind. The individual words are for the most part relatively short. One- and two-syllable words predominate. The sentences, too, are usually short, but there are many "non-sentences," that is, groups of words that have no verb but are set off as if they were dependent or independent clauses. For example, "This is the stern problem; hopeless seemingly as squaring of the circle." The adjective *hopeless* modifies *problem*, but the semicolon gives it a quasi-independent status that at first hinders our appreciation of the dependent connection. Or again, "A thing mournfully true: the stumbling-block. . . ." The grammatical construction is that of apposition, of *stumbling-block* with *thing*. The punctuation deliberately interrupts the smooth flow. The artful use of apposition is a typical Carlylean device that helps produce his effects.

Most of the words, individually, are simple and familiar; only their combination is unusual. We constantly find unexpected usage—words in relationships we would never have thought of—but when we see what Carlyle has done we realize how appropriate that usage actually is. We also find constructions that surprise us, such as the absence of a verb where we would expect a complete sentence. So, too, with the occasional rarely used word, like *palpablest* instead of the expected *most palpable*. Or the unusual verb forms like "all falsehood . . . do sink, disappearing," where we would expect "do sink and disappear," to give an obvious parallelism. Carlyle deliberately ignores parallelism, but he also avoids the confusion and faulty grammar that so often attend the lack of parallelism, as demonstrated in Chapter 3.

We compared Macaulay to a full-rigged galleon. Carlyle I would compare to a high-powered speed boat, darting here and there, twisting, bucking the waves, making rapid turns, and providing tremendous exhilaration. Nothing would be more hopeless than trying to read Carlyle by any "speed-reading" technique; the result would be chaos indeed. We must read him slowly, for we do not know what to expect. The familiar usages that permit us to absorb the sense through one glance per paragraph are simply not there.

124   Quick glances fail to grasp the significance of his prose; only by

slow reading can we appreciate the felicity of his expression.

Carlyle achieved his effects in many ways. Most of his sentences are short, but they show great variety in form and in mode of commencement. One sentence begins with an adverb, another with a conjunction, one with an imperative, another with a pronoun, another with a conjunction, another with a noun, another with an adjective. His nouns, modifiers, and verbs, generally vigorous, convey a sense of freshness. There is nothing hackneyed, nothing flabby. He avoids the monotonous doubling—the use of parallel terms to express slight differences. He does indeed occasionally use pairs, but each member contributes something quite specific. Thus, in the quotation we see *spade and mattock,* but each noun brings up an image of distinctive physical activity—the digging and swinging. Similarly with *sore heart and empty wallet,* indicating, respectively, grief and poverty—far from synonymous.

In his letters Carlyle revealed his mastery of language in a more relaxed but nonetheless effective manner. Here is his description of a visit to Coleridge [4].

> *Figure a fat, flabby, incurvated personage, at once short, rotund, and relaxed, with a watery mouth, a snuffy nose, a pair of strange brown, timid, yet earnest-looking eyes, a high tapering brow, and a great bush of grey hair; and you have some faint idea of Coleridge. He is a kind good soul, full of religion and affection and poetry and animal magnetism. . . . But there is no method in his talk: he wanders like a man sailing among many currents, whithersoever his lazy mind directs him; and what is more unpleasant, he preaches, or rather soliloquises. Hence I found him unprofitable, even tedious; but we parted very good friends. . . . I reckon him a man of great and useless genius: a strange, not at all a great man.*

What a vivid picture, achieved with such simple means! Most of the words are short and thoroughly familiar, interspersed with a few uncommon ones like *incurvated* and *snuffy.* The sentences are short, relatively, and quite uncomplicated. The force derives from the juxtaposition of common words that we would not ordinarily think of putting together. Four totally disparate nouns, when placed together, offer a remarkable picture: *religion, affection, poetry, magnetism.* The combination is powerful. So too with *unprofitable,*   125 *even tedious,* and *great and useless genius.* The combinations are superb. In general the forceful words, on which the effects depend, are nouns and adjectives.

I would summarize his style by saying that he achieves a fusion of the incongruous. He gives us the unexpected. He uses words that seemingly—by ordinary usage—do not belong together, but when he uses them we see that they are apt. They convey a precision, and at the same time a freshness that few other writers can match. Basil Willey offers a somewhat different evaluation [15].

> *Carlyle can never write urbanely; he is always on the stretch. He will not consent to use the current verbal coin; every phrase (and many actual words) must come molten from the forge. For Carlyle writes almost exclusively from the heart or the solar-plexus, not from the head. He sees by flashes and does not think connectedly; summer-lightning, not sunshine, is the light that guides him.*

With this I can largely agree. Carlyle had an inner vision and, as Willey says, transmuted "any person, scene or object, at a touch, into an emanation of the Carlylean vision."

We certainly would not want our medical writers to imitate Carlyle. What that would be like we will see in Gravenstein's parody later in this chapter. But if we study Carlyle we can gain a sense of the power that inheres in words; we can become dissatisfied with the trite way of saying things, perhaps see things more clearly and offer our descriptions or evaluations in a more effective way. We will not be afraid to express *ourselves* and avoid some of the mythologies that attend medical writing.

William Osler

Let us now turn to a medical author and a specifically medical context. William Osler was a physician of broad culture, whose essays have long been held up to the younger generation as an example of good writing. Today when we read many of these essays, we do indeed find clarity and for the most part a graceful manner of expression. Perhaps there are too many classical allusions for the modern temper, perhaps the choice of words has a certain late Victorian or Edwardian flavor, mildly pompous at times, but the overall result is pleasing.

We should not, however, offer uncritical adulation. We must realize that Osler's style definitely changed—improved—with the passing years. To illustrate this I will give three quotations from three different essays, written several years apart. The first, published in 1889, bore the title *Aequanimitas*, a quality that the good physician should cultivate. In the course of the essay Osler wrote [11],

In a true and perfect form, imperturbability is indissolubly associated with wide experience and an intimate knowledge of the varied aspects of disease. With such advantages he is so equipped that no eventuality can disturb the mental equilibrium of the physician; the possibilities are always manifest, and the course of action clear. From its very nature this precious quality is liable to be misinterpreted, and the general accusation of hardness, so often brought against the profession, has here its foundation.

The essay does indeed have a worthwhile message, but the mode of expression, of which this is a fair sample, leaves much to be desired. The striking feature is the devotion to long words of three and four syllables, and the ratio of these to the total number of words is, by modern standards, unusually high. What a mouthful we have in *imperturbability is indissolubly associated!* The whole passage reminds us of Samuel Johnson in his more rigid phases.

A few years later, in 1892, he discussed a favorite theme to which he returned again and again—the need to have clinical instruction in medical school, not to rely merely on lectures, but to bring the student as much as possible into direct contact with patients. Said Osler [13],

I would fain dwell upon many other points in the relation of the hospital to the medical school—on the necessity of ample, full and prolonged clinical instruction, and on the importance of bringing the student and the patient into close contact, not through the cloudy knowledge of the amphitheatre, but by means of the accurate, critical knowledge of the wards; on the propriety of encouraging the younger men as instructors and helpers in ward work; and on the duty of hospital physicians and surgeons to contribute to the advance of their art. . . .

Compared with the preceding excerpt this uses a simpler language, with fewer long words, and yet it still has a pompous character. 127

He does not say things simply and what he does say, important as it is, lacks grace. We realize this if we read the passage aloud. There is no sort of natural rhythm, no smooth flow. Long sentences containing many long words will rarely sound well when read aloud.

Compare this with a passage on a similar theme written almost a dozen years later. Osler was again discussing the medical school curriculum and the need to bring students into early contact with patients. This, we must remember, was written at a time when medical instruction was still largely didactic rather than clinical [12].

> *Ask any physician of twenty years' standing how he has become proficient in his art, and he will reply, by constant contact with disease; and he will add that the medicine he learned in the schools was totally different from the medicine he learned at the bedside. The graduate of a quarter of a century ago went out with little practical knowledge, which increased only as his practice increased. In what may be called the natural method of teaching the student begins with the patient, continues with the patient, and ends with the patient, using books and lectures as tools, as means to an end. The student starts, in fact, as a* practitioner. . . .

How different that is from the earlier quotations! It is clear and forceful and reads much more smoothly. The improvement, I suggest, lies in the relative simplicity of the language. Most words have only one or two syllables. There is no useless repetition, no devotion to doubling of adjectives and nouns that differ only slightly. Adjectives are used much more sparingly. There are many more verbs and verbals. In the next-to-the-last sentence the repetition of *patient* serves a rhetorical function and provides an effective emphasis.

Comparison of these three quotations will provide an insight into the various qualities that relate to style. It is as if Osler revised his own writing and deliberately set out to lighten his style. There is a change, and the change is an improvement.

A Brilliant Parody

I will close this chapter by reprinting most of a brilliant essay, published several years ago in the *Journal of the American Medical Association*. The author, Dr. J. S. Gravenstein, sent me the paper as a contribution to one of the *JAMA*'s annual Book Numbers, which I had edited for a period of ten years. The title is "New Computer Revolutionizes Writing" [6].

The following is the first report on a new computer, called Hyperbroca, which is capable of translating one English style into another. This unit will affect all writing, in science as well as in fiction. I shall present a brief history of the computer's development, a sample of what the machine can do, and a preview of the changes it might bring and the problems it might produce.

Late in 1958 the directors of the Association of University Authors (AUA) appointed a committee of seven to "study the difficulties inherent in and detracting from multiple author texts (MATs)." Each member of this committee had not only published the required 110 papers, but also contributed to at least four MATs.

Three years later the committee made its report. A single dissenting member wrote a minority report, which merely stated that no inherent difficulties exist with MATs and that careful selection of literate and competent authors guarantees excellent text. The Bible and a modern medical MAT book were cited as examples.

The majority report contains 732 pages and 1,217 references. In addition to the six committee members, 27 contributors contributed.

The most significant part was written by eight philologists who acted as consultants. On the one hand, these gentlemen considered MATs the fulfillment of all literary aspirations and they used analogies such as: "What is a single cymbalist outside an orchestra? What is a lone writer without co-authors?" and "A MAT is nothing but a literary orchestra conducted by an editor, a fabric held together by a common purpose, woven into the pattern of a higher design," and "The editor is the choreographer, the authors dance the parts," and again "The sentences march like soldiers, the paragraphs maneuver like companies, the authors make tactical decisions of battalion commanders, and the editor, the general, determines the strategy."

On the other hand, the critics complained pointedly about the uneven literary quality of MATs. They cited many examples, decried many a poor stylist, but, alas, offered no suggestion how to improve the shortcomings of MATs. One critic wailed: "Ah, could we have Hemingway describe us clearly the mysteries of RNA, could we have Joyce say beautifully but obscurely what Freud and his disciples teach! Ah, would that all professors speak with tongues of Johnson (Samuel), Swift, and Twain!"

Early in 1962 I saw this report and read the critic's cry: "Ah, would that all professors speak with tongues . . . !" At this time I was working with a team of computer specialists. Our computer translated medical texts from German into English. To be specific, into my English. Why not into that of Johnson (Samuel)? Or Swift? Or, for that matter, any style we choose? This was February 1962. Today, I can report that we now do have a computer, called Hyperbroca 1, which will translate any English text into the style of classic English authors. . . .

Once we had selected a text to be experimentally "transtylated" (from STYLE and TRANSLATE) and had chosen a number of classic English 129 *styles, we prepared vocabularies for every classic author. We then needed a dictionary of "approximate matches," correlating words and phrases in the modern text with words or phrases in the classic styles. Finally, each classic style had to be mathematically defined in the computer's language. To facilitate this task we eventually divided every style into "substyles" and found mathematical expressions for these which in turn were stored magnetically on solid state "stylets," one of several inventions coming out of this project. The stylets are inserted into the computer's first integrating unit, the Broca-1 (for Blend Readout, Override Carried Analogies). We will give all details of the computer, as soon as the patent negotiations are concluded. . . .*

For the pilot transtylation we selected authors with distinctive styles representing four centuries of English literature; for the modern English text, a well-written chapter from a recent MAT [on anesthesia]. *Rather than take paragraphs from different authors of this MAT, we decided to transtylate into the styles of several classic authors different paragraphs from the same chapter written by Vandam. . . .*

Instead of letting chance dictate a style for individual paragraphs, we asked our philologists to suggest a classic author for every section and to write a brief statement explaining their choice. This comment precedes each transtylation. [Each of the original quotations comes from an essay by Vandam, dealing with technical aspects of anesthesia.]

* * * *

The central theme of this short paragraph by Vandam is blood and equality (or lack of equality). If we wish to emphasize this theme, what tongue could render it better, what style present it more forcefully, than that of Thomas Carlyle (1795–1881), the romantic idealist, the reformer, the herald of heroes?

Original

The water/gas and blood/gas coefficients approximate each other in some instances but are divergent in others. The reason for this consists of the differing characteristics of blood in relation to the protein and lipoid content, as well as the presence of erythrocytes. The lack of homogeneity of blood, its changing characteristics from individual to individual, as well as species differences, account for the difficulty in establishing uniform partition coefficients to fit all cases.

Transtylation

And then consider how one vapor residing in gas, will here dissolve in water, there steep itself in blood, only to find the two equally thin. But not in all is blood like water! Nay, thick blood, heavy with cells and fat, may draw some vapors as sand, as dust draws dew. *That,* water never can! And blood has many faces, it lacks uniformity; in each individual, in each species it has its own countenance. With so much diversity, must we not anticipate, nay, indeed demand that one blood differ from the next in its willingness, its capacity to accept a vapor?

* * * *

Men go down. Men are afraid. This is what these paragraphs are about. That would suggest Ernest Hemingway (1898–1961). In addition the quotation from Mushin and others presents the opportunity of using dialogue. Again Hemingway is the logical choice.

Original

Among the several anesthetics, there is often a discrepancy in the various signs indicative of depth, as well as a natural variation from person to person. In interpreting these signs the beginner should bear in mind the effects of preanesthetic medication on the signs observed. He should also realize that anoxia or accumulation of carbon dioxide may likewise modify the signs.

Stage I: This has been called the stage of analgesia, or the stage of consciousness with disorientation. It terminates when consciousness is lost. Mushin, relating his own experience, speaks of an increasing disinclination to react to painful stimuli until the pain became almost unbearable. Others describe sensations of remoteness, of falling, drifting, suffocation, and visual and auditory aberration. During this stage, reassurance and suggestion help to allay fears. . . .

Transtylation

Men are not equal. Some men are put out quickly by the gas, and others fight it and reel back like boxers who have been knocked down but know that they must not stay down for the full count. If you give a man drugs before starting the gas, he goes out faster. Drugs and poor breathing change the way he reacts to the gas. Remember these things if you are a beginner.

Stage I: When the patient is awake, but feels no pain and when he does not know where he is or who he is, we say he is in Stage I.

"I felt heavy under the gas," Mushin said. "My arms were too heavy to move. I knew I had pain but I did not bother to cry out or to move," he said. Then he added, "When the pain grows awfully bad, you move after all."

"I felt far away," someone said. "And still drifting out further."

"Like falling into a manhole and losing your breath," someone else said.

"Yes, your eyes and ears can then fool you," another man said.

We must talk to the patient when he experiences all this in Stage I, and let him know that we are with him so that he is not afraid.*

* * * *

Excitement, delirium, struggling, as quickly as possible, are key phrases in this paragraph. Exaggerated responses and crass contrasts call for a theatrical style. Macaulay (1800–1859) is the choice.

Original

This is a stage of excitement, delirium, or struggling, in which consciousness is lost and there is unhibited activity. The onset may be difficult

*Note the computer's extensive use of dialog.

to detect: the entire stage may be passed through quickly if the patient has been well prepared psychologically and with appropriate preanesthetic medication. Likewise, it may be quickly transversed when the rapidly acting anesthetics are given. Breath holding, tachypnea, or hyperventilation may be encountered: the pupils may dilate. Struggling and muscular movement during this stage are among the reasons why patients should be carefully restrained before anesthesia is begun and why an attendant should be at hand during induction. Salivation, swallowing, and vomiting may appear if the patient is allowed to tarry in this stage. The chief reason for rapidly increasing the inhaled partial pressure of anesthetic is to guide the patient through this stage as quickly as possible.

Transtylation

Then comes the stage in which the patient is mad with excitement, in which he writhes in delirium, mutters without sense, and struggles without inhibition. The onset may be difficult to detect if the patient does not fall into this stage in a precipitous storm, but enters it smoothly; or if not irritating ethers but bland barbiturates are given; or if he is not trembling with fear even before the anesthetic begins, but is calm and sedate from hypnotics. And yet his heart beat may quicken, and his breath may draw fast or else stand still, and his pupils widen. For once, we have reason to ask that the physician not be alone with this patient, but that he have a stout helper; for once force, not persuasion, is required to protect the struggling patient and to prevent injury. If he tarry in this stage, the patient may salivate, he may swallow, he may vomit. The dictates of ambition, of flaunting a smooth induction of anesthesia coincide with the feelings of benevolence toward the patient and with the intentions of guiding him quickly through the dangerous straits of the second stage.

** ** ** **

Admonition and advice are here given by Vandam. Among the classic authors, Johnson (1709–1784) is readily selected as the moralist and critic.

Original

Vomiting and swallowing are no longer present, and the tone of the eyelids decreases or disappears. During this stage observation of neurologic signs and reflex responses such as touching the cornea or pinching the skin with forceps should not be used. It is better to observe the operative field and the patient's reaction to surgical stimuli.

Transtylation

He that is deeply asleep, by whatever means, desires nothing but the continuance of his sleep and is no more solicitous to vomit or swallow than to open his eyes. We do not disturb ourselves with the detection of reflex responses, which do the patient harm, and we willingly and intentionally decline to investigate the effects of touching the cornea or pinching the integument with forceps. From our notion of acceptable practice it proceeds that we discern more of the patient and his requirements if we study the surgeon's actions and the patient reactions thereto, which shew better than anything I could name the true depth of the patient's coma.

* * * *

The original text has a section heading, "Anesthetic Concentrations in Arterial Blood and in End-expired Air." It concerns the correlation between clinical appearances and laboratory measurements which are often mysterious. It's James Joyce, James Joyce, James Joyce (1882–1941).

Original

The clinical signs of anesthesia and electroencephalographic patterns have been correlated with the concentration of anesthetic determined in arterial blood. Arterial puncture and anaerobic withdrawal of a sufficient quantity of blood are followed by chemical or physical methods of analysis. During inhalation anesthesia, continuous analysis of end-expiratory or alveolar air can be performed by withdrawal through the sampling chamber of a gas analyzer, the assumption being that the alveolar gas tension approximates the arterial under normal circumstances.

Transtylation

Hoopsa, gasablood, hoopsa!
Inward the weary watcher the wavey wiggles that encephalon electrons eminate had the levels of loward lilt learn did. Pulsating pathways' puncture has spat stickly sanguis for trifles tricky troubles tested.
Send us, bright one, light one, send us quickening, quivering breath and her that weighs and it that sways to measure we take and we take and take.
Lo! The sampling chamber!
Three cheers for equality of gas in the hot bed of blood and in the chambers!
Hoopsa, gasablood, hoopsa, hoopsa!

* * * *

The reader will undoubtedly see the fabulous possibilities of transtylation. With our Hyperbroca computer we can provide an acceptable style for every author in a MAT, and we can take a single author text (SAT) and let it sparkle in the tongues of different English masters by switching styles from chapter to chapter, or, for special emphasis, from paragraph to paragraph!

References

1. Boyle, R. Certain Physiological Essays and Other Tracts. In T. Birch (Ed.), *The Works of the Honorable Robert Boyle* (London, 1772). Hildesheim: Georg Olms, 1966. Vol. 1, p. 304.

2. Browne, T. Pseudodoxia Epidemica. In G. Keynes (Ed.), *The Works of Sir Thomas Browne*. London: Faber & Faber, 1928. Vol. 2, pp. 36–37, 160.

3. Carlyle, T. *The French Revolution*. London: Chapman & Hall, 1887–1888. Vol. 1, p. 57.

4. Carlyle, T. [Letter] In B. Willey, *Nineteenth Century Studies*. New York: Columbia University Press, 1964. P. 109.

5. Glanvill, J. *Plus Ultra* (London, 1668). Gainesville, Fla.: Scholars Facsimiles & Reprints, 1958. Pp. 83–84.

6. Gravenstein, J. S. New computer revolutionizes writing. *J.A.M.A.* 204:51, 1968. (Reprinted by permission of Dr. Gravenstein and the American Medical Association.)

7. Macaulay, T.B. William Pitt. In *Critical and Miscellaneous Essays by Lord Macaulay*. Boston: Houghton-Mifflin, 1860. Vol. 6, p. 256.

8. Macaulay, T.B. Oliver Goldsmith. In *The Works of Lord Macaulay*. New York: Longmans Green, 1898. Vol. 10, p. 438.

9. Macaulay, T.B. The History of England. In G. M. Young (Ed.), *Macaulay, Prose and Poetry*. Cambridge: Harvard University Press, 1970. P. 130.

10. Macaulay, T.B. Samuel Johnson. In G. M. Young (Ed.), *Macaulay, Prose and Poetry*. Cambridge: Harvard University Press, 1970. P. 549.

11. Osler, W. Aequanimitas. In *Aequanimitas* (3rd ed.). Philadelphia: Blakiston, 1932. P. 5.

12. Osler, W. The Hospital as a College. In *Aequanimitas* (3rd ed.). Philadelphia: Blakiston, 1932. P. 315.

13. Osler, W. Teacher and Student. In *Aequanimitas* (3rd ed.). Philadelphia: Blakiston, 1932. P. 31.

14. Sprat, T. *History of the Royal Society* (London, 1667). In J.I. Cope and H.W. Jones (Eds.), St. Louis: Washington University Studies, 1959. P. 113.

15. Willey, B. *Nineteenth Century Studies*. New York: Columbia University Press, 1964. P. 104.

8

Dialects

English" is the term applied to a group of dialects that have in common a great many individual words and grammatical constructions, yet are separable one from the other through pronunciation and idioms, choice of words, and often grammar. Lifelong residents of Boston talk quite differently from lifelong residents of Brooklyn or New Orleans. I use the term *dialect* to denote any language that distinguishes one group from other groups.

Dialects owe their origin to many different factors. Geographic locale is one. Since languages evolve, the evolution will vary according to the degree of isolation. Of the many features contributing to isolation, sheer distance is important. The residents of widely separate parts of the country acquire different subcultures, reflected in the modes of speech. Natural barriers like a mountainous terrain may also induce physical separation, disproportionate to the distance in miles, and this, in turn, will induce differences in modes of speech. And even if there is geographic proximity, sociological barriers may enforce cultural separation with consequent differences in speech patterns.

Dialects often have an ethnic basis. Various ethnic groups emigrating to the United States have kept much of their original culture and much of their native speech. The immigrants have fused their native language with their adopted language to create a dialect more or less characteristic for each race. The first generation of Scandinavians or Germans or Italians, for example, created modes of expression clearly recognizable and identifiable. The second generation learned the dialect of their parents but also learned in school a different dialect that goes by various names but which I will call "standard English." The second generation of immigrants was thus bilingual like their parents. Parents and children had the first generation dialect in common, but in addition the parents remained fluent in their native language while the children became fluent in standard English.

However, dialect goes far beyond either geographic or ethnic distinctions. It also characterizes any group whose activities produce special linguistic usage. Baseball players, racing enthusiasts, stockbrokers, law-enforcement agents, rock-and-roll musicians, all have their own vocabulary and idioms—their own dialect or special mode of communication. And so too do scientists. In its broadest sense dialect is the language of a particular group, clear to the members of that group but not necessarily to members of an alien group. Various terms indicate these specialized modes of speech—slang, jargon, cant, argot, each with its own connotation—but they all are synonyms for dialect.

Slang, Jargon, and Standard English

Of these various terms I want to pay special attention to slang and jargon and their relation to the standard English taught in school. Slang has various characteristics, and no dictionary definition embodies all of them. It has the connotation of something low or vulgar, something substandard. A dictionary, if it prints slang terms, will designate them as such, to show that these words do not have full official approval.

Slang is spontaneous, vigorous, and racy, but lacks elegance. Moreover, it is ephemeral. If it becomes accepted into standard language, it then ceases to be slang—it will have improved its status; otherwise it loses its appeal and simply disappears from popular vocabulary. Slang indicates novelty—it is the leading edge of speech. Language constantly changes, and slang, having a broad popular base, embodies new usages and expressions. It creates new words or uses current words in a new sense. Thus, *nerts* is a neologism. On the other hand *lousy,* in its slang usage, invests an old word with a new context and therefore a new meaning.

I suggest one further characteristic of slang that hitherto has received little attention. Slang expresses feelings and emotions rather than conceptual precision. *Nerts* conveys the sense of total rejection but not the grounds that gave rise to the dissatisfaction. It is expressive—unequivocally so—but does not advance rational discourse in any way. Similarly, *lousy* expresses in no uncertain way the reaction of the speaker to the entity in question but does not specify reasons or identify the factors deemed faulty. An audience that has never heard a given slang word will usually have no difficulty in grasping its import, but will gain little intellectual insight into the problem that evoked the slang term. Slang expresses feelings, not ideas.

Jargon is the direct opposite—a language exquisitely precise, using terms in a highly specific sense. It is highly rational, addressed to the intellect and not the emotions; a technical language, intended for a particular group engaged in a particular activity.

In this sense jargon has special relevance to the professions. Ordinarily we think of a profession as a scholarly activity, such as law, medicine, or engineering, but this is only prejudice. Pickpockets and hoboes have their jargon along with their special skills, just as much as do psychiatrists or sociologists. Jargon has a specificity and precision of meaning, intelligible to a limited group but more or less baffling to other groups. Let me give two examples.

Suppose I declare of a particular object, "Or a lion rampant within a double tressure flory counterflory gules." Most people will be completely puzzled. Some, perhaps, will recognize the word *gules* as a heraldic term and then may realize that here we have a technical description whose meaning is only for the initiated. I doubt if any of my readers will grasp the significance of what is here written, or recognize it as a precise description of the arms of Scotland. It is indeed possible to describe the same object in words that do not include jargon, but the attempt would be enormously long-winded, would lack precision, and would repel the "in" group and the "out" group alike.

Heraldic jargon is meaningful for those who know it, meaningless for those who do not. I will offer a somewhat less exotic example from botany [3].

Leaves ovate, obliquely truncate or rarely slightly cordate at base, gradually narrowed and acuminate at apex, finely dentate with apiculate gland-tipped teeth, pubescent above when they unfold with caducous fascicled hairs, and at maturity dark green and glabrous on the upper surface, covered on the lower surface with thick, firmly attached, white or on upper branches often brownish tomentum, and usually furnished with small axillary tufts of rusty brown hairs, 3¼"–5¼" long and 2½"–2¾" wide.

This is the precise description of a particular species of linden leaf.

Compare this with a description of the same species, written for the general public and not the technical botanist—the finely toothed leaves are 3.5 to 5 inches long and 2 to 3 inches wide and are densely covered with white to brownish hairs on the lower surfaces. This gets the general idea across and permits the layman to make an   139 identification, but it does not offer the precise discrimination that a professional botanist might need. Jargon serves the professional (or the amateur who shares the expertise of the professional). It has a definite function, and only when that function is ignored should jargon be spoken of in a pejorative sense. All too often the advantageous sense of jargon is neglected.

Medicine, of course, has its own jargon, its own technical language, found most authentically in medical journals and texts. It takes its place as the dialect of a particular group, as one of the several languages we call English.

Of special importance to us is the relationship between medical jargon and another dialect that we call standard English. This, too, is the language of a particular group, but one which I find rather hard to define. We might say that the group consists primarily of schoolteachers and secondarily of pupils who have attended school. However, many who have attended school disregard completely everything taught there. Their speech and writing abound with slang, barbarisms, infractions of grammar, obscenities, and other usages condemned by teachers and textbooks alike. We could say that standard English is the dialect of dictionary-makers, or, even better, of all who have concern with something called literature (which I will leave as an undefined term).

The group that speaks standard English exerts considerable social and economic dominance. Whoever does not speak this standard dialect may find himself shut out from various social and economic opportunities. A comparable situation obtains in countries that are militantly bilingual, wherein a citizen who knows only one language will be excluded from many jobs. Similarly with standard English in this country.

Dialect and Context

The many different dialects overlap and merge into each other, even as do the groups or classes that speak these languages. Furthermore, the use of a particular dialect suggests that the speaker has some degree of kinship with the class it represents. Persons who want to establish at least a transitory kinship with other groups will try to adopt the dialect of that group. Thus, a college professor may use slang or the special argot of the young, to show that he is "one of the boys"; or talk in a sport dialect to indicate that he is not a stuffed shirt but a "regular guy." Or the movie actress with barely a high school education may feel impelled to talk about Shakespeare in the dialect of the professor. But using a dialect does not automatically bring acceptance into a group. The choice of words may be wrong, the pronunciation may be wrong—slips that mark the pretender who, like the American using high school French in Paris, is immediately spotted as an outsider.

A physician uses different dialects according to the audience he is addressing, and he adapts his language to the task at hand. He varies his vocabulary and syntax according to what he is doing, using one language when conversing with his family, another when talking to his patients, still another to his professional colleagues, and still another to his golf partners. What would be appropriate in one context would be discordant in another.

I would like to elaborate on this thought with an extended figure of speech. Since people are probably more sensitive to standards of dress than of language, I will consider language as the clothing of our thoughts and illustrate my points by reference to sartorial fashion. Two generations ago social context had a quite restrictive effect on clothes. Class distinctions were reflected in dress. The "working man" and the "business man"—the blue collar and white collar occupations—were readily distinguishable in the street car as men rode to work. The "best suit" was easily identified at church or in evening socials; and sports clothes—garments with diminished sobriety but functional advantage—were suitable only in quite restricted circumstances.

Today the variation in clothes is dazzling. Standards are relaxed, informality is widespread. Men will wear leisure clothes where they formerly dressed up. We see sports clothes in church, blue jeans and sweaters at the opera, while restaurants invite customers to "come as you are." Yet it is not quite true that "anything goes." For example, some restaurants are explicit on proprieties of dress (and the need to post a notice is in itself an interesting phenomenon), and many businesses and institutions require standards of attire in their employees. Moreover, whoever receives an invitation with the notation "black tie" will not wear blue jeans. Many situations demand a conformity, enforced by various sanctions.

If we compare clothing to the various dialects that a physician uses, we can see how his choice will depend on various factors, comparable to those that influence his choice of clothes. There is fashion, there is convenience and comfort—we might call it suitability, and there is compulsion.

In the linguistic habits of physicians, I will distinguish two major contexts, social and professional. In the first the physician is a member of the community at large—a family man who shares in the concerns of society along with other members of the larger community. In this context he uses variations in the dialect of standard English according to his general education, cultural habits, and linguistic skill. Although these will obviously not be equal among different members of the class, I would disregard the differences and settle on a mythical average. A further distinction I would also disregard, namely, the difference between oral and written communication. Some individuals are much better at one than at the other. But here again I consider only the average.

The second context involves professional activity wherein the physician uses technical language quite different from standard English. Here, too, there will be considerable variation both in writing and in speech, but we must disregard individual differences and think of averages.

In each of these contexts a historical perspective will reveal marked changes, comparable to the changes in dress. If we go back 150 to 200 years, we find that the standard English of the day and the technical medical language were much closer together than now. For the most part a medical text was readily intelligible to the educated laymen, for the technical language of medicine—or science generally—was not as well developed as today. Most of the words in medicine enjoyed a reasonably familiar usage, and those with technical meaning were only mildly esoteric. The medical curriculum occupied much less time than it does today. We should also note that literary English—the higher reaches of standard English—was more complex.

Within a century and a half, certain pronounced changes took place. Standard English (like fashions in dress) became simplified, sentences shorter and less turgid. There were fewer modifiers and subordinate clauses, less redundancy, fewer abstract terms. At the same time, the technical language grew more complex and difficult. Progress in various sciences had led to a greater precision and more exact discrimination that in turn required a more precise nomenclature to identify specific meanings. Medical language diverged more and more from standard English. As a corollary, students had to spend more time studying medicine so that they could become more adept in the special vocabulary; and they had to have more preliminary education in the sciences before they could even approach the special vocabulary in a meaningful way. This was a hidden cost of medical progress.

However, the medical dialect involves more than vocabulary. It expresses certain characteristics of the class that uses that dialect. The class of physicians, closely allied to the class of scientists, participates in the ideals of that class, and in its jargon reflects the (alleged) properties of the scientist. Language aside, how does the scientist differ from other people in society? Without trying to answer fully this difficult question, I suggest the following attributes as especially relevant to our discussion. The scientist, among his other characteristics, is objective, impartial, and accurate. Ideally, he aims at mathematical expression; his objectivity requires that he erase any subjective bias and deal only with "facts," presented in such a way that any competent person can repeat the observations if he adheres to the same procedure.

Hence arise certain conventions. To be objective, a scientist must not intrude the first person into his writings. The passive voice seems to eliminate subjective bias and therefore should be cultivated. Colorful language, since it introduces the subjectivity of an observer, must be avoided. Quantitative expression, as the language of science, is at all times preferable to qualitative description. In anything but the simplest case report, data must have statistical validity.

These beliefs (regardless of their validity), when carried over into medicine, helped to determine the form of medical dialect, that is, the selection of words and the modes of putting them together. As scientific language grew in importance, the medical dialect expanded.

We see this quite forcefully if we refer to the medical journals published just before the turn of the century and compare them with the journals of today.

Furthermore, as the medical dialect became more and more set in its ways, it took on a symbolic status. Whoever used it became identified as the member of a class, and the status of the class tinged everyone who used that language. "Scientific!"—a word of almost magical value that most physicians eagerly welcomed.

But language never stands still. At any given moment there are the conservatives who want to reject any further alteration and preserve the existing format, and the progressives who welcome further change and may even want to alter its direction. Vested interests appear, who seek to prevent lapses from the existing dialect. Most editors of medical journals, along with most contributors, belong to the conservative wing. To continue my comparison of dialects with fashions in dress, I would say that these editors resemble business executives who demand that their employees wear shirts with starched collars and suits with vests. The executives have the power to enforce their demands. So too with editors, who in demanding a certain linguistic pattern from their contributors have also the power of enforcement, namely, the power of rejection. On the other hand, the editors and contributors who want greater freedom in writing are, in terms of the metaphor, seeking permission to wear leisure jackets and sport shirts.

To what extent should a medical dialect be kept distinct from standard literary English? Can medical dialect have a high degree of overlap with standard English, so that medical language is distinguished by a technical vocabulary but otherwise has the same values? The unsophisticated reader may, in his innocence, say, "But of course!" Let me indicate some of the difficulties.

Not infrequently the contributors to medical journals complain bitterly about the changes inflicted upon the manuscripts. One author wrote, "Why do most journals and publishers employ a collection of copy editors, proofreaders, and the like who take what is a perfectly acceptable and clear text and effect a series of minor changes that usually add nothing and occasionally obscure the meaning? . . . Why must these obnoxious obscurantists be allowed to convert an interesting, well written and lively paper into an emasculated neutralized characterless tract?'' [2].

One answer is that the copy editors are following the regulations of the chief editor, on whom all responsibility must ultimately fall. If he chooses to delegate the power to some subordinate, he must nevertheless bear responsibility for any dull or wooden quality of the prose that his journal prints.

We cannot help sympathizing with the author of the letter, and yet we must also defend the copy editors. Authors have surprisingly different notions of spelling, punctuation, abbreviations, bibliographic format, and the like. One author may speak of *pediatrics,* another of *paediatrics;* one will refer to a unit of volume as a cubic centimeter, another as a milliliter; one will number his references, another arrange them alphabetically. Furthermore, manuscripts will usually lack consistency in any given respect, and in addition will have errors and omissions that the author did not catch. To make manuscripts internally consistent and conform to some degree of standardization requires a great deal of work. This devolves on the copy editors, who must go over abbreviations and spelling; bring uniformity to the references; check figures, tables, and diagrams; provide the marks necessary to communicate with the printer; eliminate grammatical errors; and perform comparable duties. These tasks require diligence and care.

There is, however, a further task. Most authors, in addition to being careless, lack writing skill. The copy editor who oversees the consistency of details also has the responsibility for correcting faulty or obscure English, getting rid of ambiguities, achieving clarity, and making the language conform to reasonably literate standards. And here we run into trouble, often severe trouble.

The copy editor is not a scientist, nor does he come to the job trained in medical dialect. A good copy editor will have an excellent command of standard English and a good ear for this dialect but will need to learn the technical language of medicine. Then he will face the dilemma, When the text needs revision, into which form should it be put? The medical dialect is often unnecessarily technical, may often run counter to pleasing standard English. What to do, when the original text sounds a note discordant to a sensitive ear? The copy editor, in trying to eliminate sheer bad writing, may inadvertently use terms that distort the meaning of the author, and the latter, not unjustifiably, may feel aggrieved.

The letter quoted above does indeed express a legitimate grievance but it does not take into account the dreadful writing that disfigures so many manuscripts. One great difficulty is the custom, in the United States, of leaving the correction of bad writing to a copy editor, whereas it should devolve on a senior editor who is medically competent *and* who has a good ear for standard English. We commonly hear that articles published in the British medical journals are much better written than those published on this side of the Atlantic. Such a complaint does not take into account the possibility that the British papers are better edited by medically trained personnel.

Dependence on Rules and Numbers

One "solution" prevalent in American medical publishing is the establishment of rules—prescribed usages in regard to choice of words and syntax. In theory the rules will promote clarity and eliminate ambiguity, but while teaching rules is easy, we cannot teach when to apply the rules. This is a matter of judgment. For example, one journal will not permit the word *doctor* to serve as a synonym for physician, since a Ph.D. is as much a doctor as an M.D., and *doctor,* unqualified, does not tell us which degree is involved. Imagine the consternation of an author when, in his discussion of Chinese medicine, the term *barefoot doctor* had become transformed throughout to *barefoot physician.* The copy editor was carrying out the precepts of the rule book but did not use judgment.

Again, in some journals an author may not say, "In patients of over 40," since this does not show the unit of measurement. The text must read, "In patients over 40 years of age," as if to rule out the possibility that the author meant "over 40 centimeters in height." Editorial changes such as these would justifiably irritate the author of the letter quoted above.

I have only praise for a good copy editor. But in medical periodicals good copy editing is rare, so that the published article merely enshrines bad writing. Clumsy language, published in a "good" journal, encourages further writing of the same sort; when a senior

physician writes in this fashion, and his writing appears thus in print, he is encouraging his juniors to fall into the same pattern. Teachers in medical school, furthermore, impose conventions on their students who, once infected, rarely recover. Sometimes, however, there is rebellion. One young physician related the following experience. In college he had had the ability to express himself well, only to become frustrated in medical school which emphasized "dry, colorless language." He mentioned writing a report wherein he described his experiments on enzyme kinetics, which involved a vast amount of data. Because he described the experiments in terms like "gargantuan" and "tedious," his professor told him that his biochemistry paper "was best fit for the readership of the *Ladies Home Journal*" [1]. Undoubtedly, the professor thought he was making a devastating comment; actually, he was merely demonstrating the blight that affects medical literature.

The difficulty, I suggest, centers around the false worship of quantitation. Ideally, science expresses itself mathematically, whereby unequivocal symbols can have meaning, and words from ordinary language are unnecessary. However, since ordinary words are at present indispensable, the idolators of quantitation insist that the words be as neutral as possible, that the physician approach the mathematical ideal as closely as he can. Hence the writer must use terms as neutral as possible and avoid any expression that might offer the slightest ground for ambiguity.

This view I believe to be philosophically unsound and subject to all the pitfalls of extreme reductionism. Since I cannot here argue the case in detail, I will content myself with throwing a little fuel on the fire, to arouse controversy rather than settle anything.

The passion for quantitation often leads to illusion rather than precision, and to impoverishment of language rather than accuracy. Think for a moment of the old way of reporting autopsies, when description relied on comparison and metaphor. How often have we heard modern ridicule for the old pathologists who compared lesions to foods or other indifferent objects—a concretion like a grape seed, a tumor the size of a pea or perhaps a grapefruit, an inflammation like a purple grape, or a mass the size of a tennis ball. Although tennis balls have a uniform size, fruits do not. Today reference to foods or other familiar objects is not quantitative enough, and therefore the prosector must measure a lesion or organ in centi- 147 meters and provide some numbers that supposedly yield precision and objectivity. I wonder whether the coefficient of error, as a prosector goes with a celluloid ruler from one lesion to another, is any less than the discrepancies we find if we compare lesions with green peas; and I wonder whether a simple comparison like "small grapefruit" may not be more illuminating than a series of figures. I would also ask, How much of the quantitation has any real point? and whether we ought to cultivate judgment in our students so that they will recognize *when* precision is important, rather than inculcate a blind worship of figures for their own sake. I believe medicine would be much better off if we replaced much of our would-be quantitation with expressive qualitative words that reflect awareness of the world around us. Relevance and comparison can be more meaningful than figures. Quantitation can be important, obviously, but I deny that it is more important than the judicious use of language. And standard English is a magnificent language.

Too often medical writing is stilted and artificial, noun-oriented, abounding in passive constructions and excessive prepositions, clumsy and repellent. Under the cloak of "science" these defects get reevaluated as positive merits; soon they are defended by vested interests like the professor who emasculated the student's report, and the editors who value pedantry over clarity and simplicity.

Medicine does indeed require a technical language, just as do heraldry and botany; but that language need not be incompatible with good standard English. There is no inner necessity for a technical dialect to go hand in hand with bad writing. And there is no need to mistake obscurity of expression for profundity of thought.

References

1. Cramer, S. F.   Letter to the editor. *N. Engl. J. Med.* 294: 1242, 1976.
2. Morgan, W. K. C.   Verbal blemishes seen as virtues (letter to the editor). *N. Engl. J. Med.* 287: 941, 1972.
3. Sargent, C. S.   *Manual of the Trees of North America* (2nd ed.). New York: Dover, 1965. Vol. 2, p. 745.

9

Synonym, Context, and Translation

English draws its words from many sources, in particular from the Greek, Latin, and Romance languages on the one hand and from the Teutonic languages on the other. All have contributed to the richness of modern English as a medium of expression.

The user of English has at his disposal a wealth of synonyms, among which he can find alternative ways of expressing a given idea. Yet no two words will have meanings the same in all respects; somewhere along the line they diverge; synonyms provide a similarity in some respects but not in others. As the *Oxford English Dictionary* indicates, synonyms have "the same general sense," but each of them has meanings not shared by the others and exhibits different shades of meaning in different contexts. Under certain circumstances the meanings overlap; in others they diverge. The *O.E.D.* gives an example from Prescott's *Philip II*: "The name of soldier is synonymous with that of marauder." Clearly, there are many soldiers who are not marauders and many marauders who are not soldiers. However, during the historical era of Philip II, soldiers behaved outrageously and plundered wantonly. Under those circumstances the meanings of *soldier* and *marauder* overlapped, but in other contexts, other circumstances, they did not.

Context means literally a weaving together. It implies a union of parts into a whole, whereby the meaning of the whole determines the meaning of an individual part. As an illustration of this we might think of words with the same spelling but different pronunciations. The sentence as a whole tells us which meaning is appropriate. Thus, *lead* in one context is a noun indicating a particular heavy metal; in another it is a verb, meaning "to conduct" or "to guide." *Let* in one context means "to permit"; in another it describes a feature in tennis wherein the serve has been hindered and the point must be played again.

Words are synonyms only when there is a partial overlap of contexts. In many situations, for example, *belly* and *abdomen* are synonyms. "The patient had a pain in the belly" and "The patient had a pain in the abdomen" have the same medical significance. But in a different usage *belly* and *abdomen* are no longer synonyms. We might speak of the belly of a sail or a violin, but not the abdomen of a sail or a violin. And abdomen dance is not interchangeable with belly dance.

Through usage words extend their meanings into different con- 151 texts, get settled there, and take on a new sense, often metaphorical. Think, for example, of the modern, more or less slang, significance of *square* or *camp.*

Words from different roots that at one time had similar meanings may subsequently diverge. Take, for instance, *gut* and *intestine. Gut,* of Teutonic origin, meant the contents of the abdominal cavity, the entrails. Then the word became directed to a particular entrail, the lower alimentary tract. But by extension it came to have other meanings, including violin strings, the silken fiber from the intestine of a silkworm, and a narrow channel or passage, as, a channel of water. On the other hand, *intestine* comes from the Latin *intus,* meaning *within.* It referred to the lower part of the alimentary canal but did not acquire a host of accessory meanings. *Gut* and *intestine* are most certainly synonyms for the lower alimentary tract, but the congruence of meaning stops there.

We might think of synonyms as intersecting circles. The portions that intersect represent the identity of meanings, and in that area the words are interchangeable. The portions of the circles that do not intersect represent the meanings unique for each, not applicable to other contexts. We might think of each circle as having a flavor, an indefinite aura, a composite imagery derived from all the varied usages. The flavor of the whole affects that part of the circle acting as synonym. The word *square,* for example, applied to the highly conservative person out of touch with modern trends, conveys a definite image. The word calls up the properties of a geometric figure—something stable, with sharp angles. The person we call square seems to take on these properties as we use the word. Different synonyms will each bring their own metaphorical flavor that affects the total prose picture.

Synonym and context are essential terms in translation. When we learn a foreign language we start with the rudiments of grammar and an elementary vocabulary. We have lists of words in the one language and a list of equivalents in the other. The two lists represent synonyms. Then, after gaining a little overall familiarity, we start on an elementary reader. Those to which I have been exposed

always had a vocabulary in the back, giving in alphabetical order the foreign words, each with its English equivalent.

The beginning student may get the idea of a one-to-one equivalence—only one right English word for each foreign word. He ignores the concept that synonyms—equivalence—depend entirely on context. And if you find a different context, you cannot assume that the two words, the foreign and the English, are still equivalent.

A reverse example of this I encountered in a small French hotel whose owner had a limited knowledge of English. Prominently displayed in the room was a sign in English that began, "The direction does not answer for things of value. . . ." The source of his difficulty is easy to find. The French *direction* does indeed correspond to *direction* in English, but only in a particular context. For the sense that the owner intended, the proper translation of *direction* would be *management.* Similarly, the word *répondre* does indeed sometimes mean *to answer,* but in a different context it can mean *to be responsible for.* What the owner had said to himself in French would be rendered in English as "The management is not responsible for valuables. . . ." We cannot translate one language into another without thoroughly understanding what word a particular context requires in order to make sense.

The situation lends itself to an amusing game. Take a sentence in a foreign language, look up every word in a good dictionary, and then, while preserving the grammatical structure, use in your translation English words found in the dictionary but intended for a different usage. Such a disregard of context results in amusing nonsense. Here is an example taken from La Mettrie's *L'homme machine.*

From what I have just said the best society for a husband of inspiration is his own, if he does not detect a similar one. The ghost becomes rusty with those who do not have one, for want of being drilled: in tennis one badly discharges the bullet at one who badly attends it. I would prefer a sharp soldier who would not have had any breeding than if he had had a bad one, provided he was still immature enough.

This, I am sure, baffles the understanding. The "proper" translation [1] is

From what I have just said, it follows that a brilliant man is his own best company, unless he can find other company of the same sort. In the society of the unintelligent, the mind grows rusty for lack of exercise, as 153

at tennis a ball that is served badly is badly returned. I should prefer an intelligent man without an education, if he were still young enough, to a man badly educated.

The original French, from which this translation is taken, reads [1, p. 25]

Ce que je viens de dire prouve que la meilleure compagnie pour un homme d'esprit, est la sienne, s'il n'en trouve une semblable. L'esprit se rouille avec ceux qui n'en ont point, faute d'être exercé: à la paume, on renvoie mal la balle à qui la sert mal. J'aimerais mieux un homme intelligent, qui n'aurait eu aucune education, que s'il en eût une mauvaise, pourvu qu'il fût encore assez jeune.

If my translation were intended seriously, it would be considered atrocious. Unfortunately, some translations of scholarly works, seriously offered to the public, are almost as bad, show a profound disregard of context, and yield a result that at best is confusing, at worst utterly wrong.

Here, as an example, is a published translation from the German [4]. The subject matter is the so-called healing power of nature and the changes that this concept had undergone.

Here the history of the conception of nature and its numerous changes cannot be considered in detail, only it should be said that the representation of nature as an almost personal, consciously purposeful managing nature, standing above the material, no longer seemed maintainable since Galileo, Bacon, and Cartesius, rebuilding anew on the basis of discovery, had overthrown the scholastic-Aristotelian world conception.

The reader may get a rough impression of what the author intended but certainly not a precise account of the original ideas. At best the wording is clumsy, with less than optimal rendition of particular German words, and in two places the translation is wrong. Before offering a better translation I give the original German text [3].

Es kann hier nicht auf die Geschichte des Naturbegriffes und seine mannigfachen Wandlungen eingegangen werden, nur das sei gesagt, dass die Vorstellung der Natur als eines beinahe persönlich gedachten, über der Materie stehenden, bewusst zweckmässig handelnden Wesens nicht mehr haltbar schien, seitdem Galilei, Bacon, Cartesius von Grund auf die Erkenntnis neu aufgebaut, den Sturz der Scholastisch-aristotelischen Weltanschauung herbeigeführt hatten.

The word *material* in the published translation is wrong. It should be *Matter,* referring specifically to the Aristotelian conception. Then, "rebuilding anew on the basis of discovery," also wrong, misconstrues the German words. Furthermore, even where the rendition is technically not incorrect, it is awkward and confusing. For comparison with the published version I offer a different translation, with a more precise choice of English words, better adapted to the context of the original.

We cannot here take up the history of the Concept of Nature and its many transformations. Let us say only that the idea of Nature, regarded as an almost personalized Essence standing above Matter, conscious, and acting in a purposeful manner, appears no longer tenable, ever since Galileo, Bacon, and Descartes, on the basis of new-found knowledge, had brought about the collapse of the scholastic and Aristotelian world view.

The principle of translation is to stay as close to the original sense as possible, and yet produce idiomatic English. The translator must choose a word that reflects the original meaning, fits into the context, and is completely idiomatic.

I could give many instances of thoroughly bad translation, so bad, often, as to be incomprehensible. While examples abound today, the defects were much more widespread in earlier times. In the past, especially in the seventeenth and eighteenth centuries, translation was a drudge job, a means whereby an author could eke out a bare subsistence and provide himself with an income that, however meager, would still permit him to write his play or his poems. A publisher might sniff out a market for an English version of some continental work and then find someone to provide the translation. Grub Street was the term to describe not so much a physical locus as a general activity. Publishers would commission poverty-stricken authors to translate various foreign works at page rates. There was

155

no standard of excellence to be met, no critical evaluation of the manuscript before it was printed.

The drudge who did the work might or might not be knowledgeable in the particular field. He would have, say, a knowledge of Latin or French, sufficient to turn out a rapid translation that bore some relationship to the original. Speed was of the essence, since payment was by the page. With this method of payment, and with no critical examination of the result, the translator had no incentive to make an accurate rendition.

As a medical historian I have had occasion to study intensively the writings of seventeenth- and eighteenth-century physicians. Most of them wrote in Latin. Many of the major texts had been translated into English, but a modern scholar needs only a brief exposure to these translations to appreciate their frequent obscurity. With some exceptions the available English renditions gave a vague (even though fine-sounding) exposition that might serve for generalities but have little value for rigorous analysis. For the most part the English editions served only as "finders," to give a general sense that this or that passage was probably important. To appreciate the fine points the reader would have to go to the original Latin. Of the great medical classics of that era, originally written in Latin, only a small proportion had a satisfactory version in English.

If we try to analyze why the translations as a class are so bad, we can note several causal factors. First, many of the translators were not competent in the particular area of scholarship. A hack writer well versed in Juvenal or Lucan, would not, a priori, be expected to understand the subtleties of the mechanical philosophy or its application to physiology. To produce a good translation a writer must have a sound knowledge of the subject matter involved. Many translators of medical works, for example, were neither physicians nor scientists; they could not grasp subtleties of expression or fine points of doctrine.

But even assuming that a translator has a thorough command of the subject matter in question, he also needs a mastery of the English language. He must recognize the shades of meaning that the original terms involve, and then express these in English prose that will convey the original distinction. Indeed, it is more important for a translator to be a master of the language into which the work is to be rendered

than a master of the original language.

As a third requisite I suggest that the translator must avoid haste. Sometimes a writer can dash off an original composition at white heat and record his inspiration while it is still glowing in his mind. Translation, however, can never be accomplished in passion. It requires deliberation, scrutiny, and questioning: Is this word or that word best under the circumstances? A decision may require considerable thought and time.

I will give a further example from the German. The context has to do with the healing of wounds and the relation of this process to the more abstract "healing power of nature." The published translation reads [4, p. 50], "The healing of wounds, the substitution of defects occur through the same process which effects nutrition . . ." The German word rendered as *substitution* is *Ersatz.* In certain contexts this is a correct rendition. For example, during the war, when there was a shortage of food, we frequently heard of *Ersatz*—substitute. But when the context is pathology and the healing of wounds, then the phrase "the substitution of defects" makes no sense. The context indicates a process of repair or restitution or restoration. In German dictionaries I was unable to find any translation of *Ersatz* as *restitution,* although I did find *replacement.* But the failure of a dictionary to include the word I consider appropriate does not bother me. In this context I translate *Ersatz* as *restitution* and I will let the dictionary catch up with me in some future edition. I will not mangle English to fit a dictionary that has never encountered this particular context. My own translation would read, "The healing of wounds and the restitution of defects takes place through the same process that induces nutrition."

The translated sentence, the first part of which is given above, continues,

> . . . and also these processes did not utilize favorably the healing power of nature in the usual sense, because also tumors, monsters, etc., in the same way require growth from nature, and indeed are non-purposeful or indeed even lethal to the organism.

I offer a translation which is better English (as well as being closer to the original text):

> *But we cannot consider these processes as favorable to the healing power of nature in the usual sense, for tumors and malformations and the like are brought about through the same manner of natural growth, although they serve no purpose for the organism or may even be destructive.*

This, while thoroughly grammatical, could still be improved. A freer translation, however, would be less faithful to the original. Here we have the difficult problem, How close to the original text should a translation be?

Quite commonly this question takes the form, How *literal* should a translation be? To what degree should there be a word-for-word rendition? The word *literal* embodies gradations depending on context and purpose. The *idiom*, for example, is a rock on which all translations can founder. Take the simple query, "How do you do?" To attempt a word-for-word translation into another language would be nonsense. Yet most languages have an equivalent question, with an identical context and purpose. In translating an idiom we try to give an equivalent, and an equivalent is by no means a literal rendition.

Differences in grammar offer a further insurmountable barrier to literal translation. Think, for a moment, of the reflexive constructions so common in languages other than English, or—an example I have used before—of the long chains of modifiers that can precede a noun in German. These modifiers may themselves have modifiers and qualifications that would make any word-for-word rendition intolerable. Again we must fall back on the notion of equivalence—getting hold of the sense and rendering that sense in a form that will be idiomatic, properly grammatical, and yet close to the original.

Most students of Latin have had experience with the interlinear translation, more commonly known, perhaps, as a pony or trot. This is the closest approximation to a literal translation. Above each Latin word is an English word in translation, but the order of the words remains Latin. To offer a respectable translation the student must rearrange the words into an order appropriate for English.

We can contrast the interlinear translation of Caesar or Cicero with the scholarly texts of the Loeb Library, which offer the original Latin (or Greek) on one page and the English translation on the facing page. These scholarly renditions do not try to be literal but they are remarkably close. They provide an equivalent in English for what the author had said in the original language.

To be sure, in a few simple examples the equivalent and the literal will coincide. "The pen of my aunt is black" can be translated literally into almost any foreign language, and sentences of this type are useful as introductory exercises. But the literal and the equivalent soon diverge when we progress into more complex speech or into what we may properly call literature. If we are translating poetry, or drama, or a novel, or a scholarly treatise, or a political speech, or a news report, what does *equivalent* mean? Each of these modes of language has a different purpose, a different context, and these will affect the character of the translation.

How, for example, should we translate poetry? Should the translation try to preserve the rhyme scheme of the original, or forego rhyme while still remaining metrical? In either case, should the translation preserve the original meter? Or should the rendition be in prose? To illustrate variations in the concept of *equivalence,* we can think of the numerous translations of the *Iliad* or the *Inferno*, Goethe's *Faust* or Molière's verse comedies.

Complete equivalence in all respects is impossible. Something must give. One translator, wanting to keep the original rhyme scheme, will diverge from what the author actually said. Another translator, who wants to keep closer to the original text, will give up the rhyme and translate into free verse or prose. In any case the translator will claim that he keeps to the spirit of the original. But *spirit*, a term of great latitude, will mean different things to different people, and therefore different translations of the same poem will vary tremendously.

Translating a work of fiction offers problems somewhat less severe than those of translating poetry but still comparable. Although there is less emphasis on imagery, a particular locale, peculiarities of social behavior and adaptation, and colloquial speech can play an important role in narration. The translator must convert the idioms of one language into those of another: He must render the situations and behavior patterns meaningful for an audience that may have a different culture. What to do, for example, with patterns of speech and behavior, immediately understood by the original audience but perhaps rather obscure for readers far removed in time and space? These problems, especially acute in translating fiction, may also be troublesome in translating expository prose.

An unskilled translator may use vogue terms that enjoy a brief popularity but which, to a later generation, will seem stilted and even ludicrous. Such a translator has tried too hard to use idioms of the new language to express the ideas and actions of the old. Yet if he sticks too closely to the original words, his effort may seem wooden, lifeless. A good translation will always diverge from any literal rendition; but even though to some extent free, it will nevertheless be *close.* We cannot reduce the degree of freedom to a formula or the *closeness* to a mathematical percentage. Good translations are works of art. And good translations are rare.

In works of nonfiction such as an essay the author is trying to convey concepts, whether in history, science, philosophy, literary criticism, economics, or other discipline. Here a translator must have a maximum degree of precision. He must indeed catch the spirit of the original, but *spirit* has a significance somewhat different from that exhibited in a poem or novel. An essayist deals with ideas, which are fragile. Their validity and meaning depend on the underlying background, and the spirit that the translator must catch relates to the entire cultural and intellectual framework on which the ideas rest. To be sure, no two persons will have the same insight into the culture of a period. Since the degree and character of the insight forms the basis of any interpretation, translators will offer differing interpretations. Such a difference is just as inescapable as the variation between individuals.

The translator, then, thoroughly familiar with the background of the subject with which the essay deals, must render the ideas in the essay into a different language and do so with the utmost possible precision. Insertions must be at an absolute minimum, limited, chiefly, to situations wherein a single word in the foreign language may require several English words to convey the sense. Similarly, the translator must not make deletions from the original unless, again, several of the foreign words can be rendered by a single English word.

I will give two different translations of a Latin passage, originally written by Francis Bacon early in the seventeenth century. Bacon was dealing with what we would now call scientific method and the establishment of scientific truth. He emphasized the collection of data, but insisted that the data be reliable. The philosopher, David Hume, writing in the early eighteenth century, translated as follows (I quote only the first two sentences) [2]:

> *We ought to make a collection or particular history of all monsters and prodigious births or productions, and in a word of every thing new, rare, and extraordinary in nature. But this must be done with the most severe scrutiny, lest we depart from truth.*

Hume was an excellent English stylist and his choice of words is admirable. Yet we have no difficulty in detecting a definite eighteenth-century flavor.

The most widely used translation of Bacon today is that by Spedding, Ellis, and Heath. They offer the following translation of the passage in question [6]:

> *For we have to make a collection or particular natural history of all prodigies and monstrous births of nature; of everything in short that is in nature new, rare, and unusual. This must be done however with the strictest scrutiny, that fidelity may be ensured.*

The sense is the same, but the flavor is different.

Let us now look at the original Latin [5]:

> *Facienda enim est congeries sive historia naturalis particularis omnium monstrorum et partuum naturae prodigiosorum; omnis denique novitatis et raritatis et inconsueti in natura. Hoc vero faciendum est cum severissimo delectu, ut constet fides.*

We can see how each translator offered a slightly different rendition of the Latin terms. All the translations are close to the original but they exhibit different shadings.

A reading knowledge of one or more foreign languages is widespread among educated Americans, very few of whom, however, will ever have occasion to publish a translation into English. Yet a close attention to problems of translation will be extremely helpful to those who want to improve their skills in English composition. Merely comparing a published translation with the foreign original is valuable, but the truly rewarding exercise is to take two different translations of the same foreign text and compare them with each other and with the original. The two English renditions will certainly diverge—sometimes to such a degree that they scarcely seem to be rendering the same passage. But even if the two are fairly close and the differences between them subtle, a student will gain vast critical appreciation if he will study those differences, weigh each word in relation to the context, and judge for himself which translation is preferable—and why. He may even find that neither of them is really satisfactory.

The effort involved in all this may be considerable, but the reward will be a sharpened appreciation of context. This single word can epitomize all the problems of writing: *What is the right word in this context?* The right word is the one that gives that sense of aesthetic fit that I noted earlier, the satisfaction that marks the successful completion of an artistic effort. Exercises of this type will help the student appreciate the value of words in relation to context. And this, I suggest, is the central problem in the art of writing.

References

1. de la Mettrie, J. O. *Man A Machine.* La Salle, Ill.: Open Court Publishing, 1943. Pp. 25, 97.
2. Hume, D. *Enquiries Concerning the Human Understanding* (L. A. Selby-Bigge, Ed., 2nd ed.). Oxford: Clarendon Press, 1951. P. 129.
3. Neuburger, M. *Die Lehre von der Heilkraft der Natur im Wandel der Zeiten.* Stuttgart: Enke, 1926. Pp. 49, 53–54.
4. Neuburger, M. *The Doctrine of the Healing Power of Nature Throughout the Course of Time* (translated by L.J. Boyd). New York, n.d. Pp. 46, 50.
5. Spedding, J., Ellis, R.L., and Heath, D. D. (Eds.). *The Works of Francis Bacon* (London, 1857–1874). Facsimile edition, Stuttgart: Fromann, 1962–1963. Vol. 1, p. 283.
6. Spedding, J., Ellis, R.L., and Heath, D.D. (Eds.). *The Works of Francis Bacon* (London, 1857–1874). Facsimile edition, Stuttgart: Fromann, 1962–1963. Vol. 4, p. 169.

10

The Book Review

The Purpose of Book Reviews

What is the function of book reviews? Why do they exist at all? How does a review differ from an advertisement? These are all meaningful questions that deserve specific answers.

As far as professional groups are concerned, there are five interested groups: the authors who write the books; the publishers that publish them; the public that reads them; the libraries and individuals that buy them; and the journals that, in one or another way, help to bring the other groups together. Within a profession, authors want to establish a reputation; publishers are in business to make money; readers seek information and try to keep up with important developments, and may buy books (or get them from the libraries that buy them); and professional journals render service to the readers. Where do book reviews fit into all this?

Publishers in the aggregate issue thousands of books each year and want to promote sales. A publisher must bring books to the attention of prospective purchasers. The audience for professional books is relatively limited, and the promotional devices appropriate for popular books—like television interviews with authors—do not apply. To bring books to the attention of readers in the professions, publishers rely partly on notices and reviews in suitable journals.

A professional journal exists to keep its readers informed on the latest research efforts and developments. A journal offers articles, essays, and various communications that may be important for its readers who want to keep up with what is new. Hence, journals, to perform their function, must inform their readers about relevant new books. Book reviews and notices are thus a service to the readers. All this sounds obvious when regarded quickly and abstractly, but when we approach concrete particulars we find certain difficulties.

We must appreciate the intense competition that goes on at all levels. I will speak principally from the standpoint of the journals, and stress the competition for space. To start with, the editor-in-chief must decide how much of the journal will be given over to the papers describing original research and new observations, how much to letters and editorials, to news and announcements of various kinds, and to book reviews. Each editor will decide for his own journal. We cannot here go into the factors that influence his decision, but let us merely accept the result, that only a limited number of pages are available for the book reviews. **165**

The publishers, who want to bring their products to the attention of potential purchasers, compete for this space and are eager to get notices and reviews in journals of wide circulation. Such journals receive vastly more books than they can possibly review. Last year, for example, in the office of the *JAMA,* I received over 2,200 books of all categories, and only about 20 percent of these were reviewed. The well-established publishing firms that specialize in, say, medical books, routinely send books to the major journals in the field. Less well established firms, particularly, want to get their books reviewed in suitable journals. An editor may be bombarded with publishers' releases, bringing to his attention forthcoming books, telling him how important they are, how much they will interest his readers. Often the release will include a capsule review made up in advance, so that the editor, if he is so minded, can create an instant review without bothering to read the book. New and insecure publishers are especially pushy in their releases, which may often repel a fastidious editor.

The editor, facing the enormous numbers of books that cascade endlessly into his office, must decide what to do with them. Their variety is indeed startling. A medical journal obviously has special responsibility for those works that pertain to medicine, but how to define medicine, determine its limits, and decide which books are relevant? Medicine includes dozens of clinical specialties and subspecialties, and the numerous basic sciences and their subspecialties, all of which have their own texts and monographs. There are, as well, texts written not so much for physicians as for accessory personnel—nurses, technologists, physicians' assistants, and other members of the health team. Then there are books on the sociology or economics or history or philosophy of medicine, or medicine in relation to government and governmental policies. There are books designed particularly for the mature physician and those intended primarily for students; there are books addressed to the layman to keep him informed on various diseases or on scientific advances; there are books in the muckraking category, and those by obvious cranks with an axe to grind. There are novels with a medical theme, and others that deal with medicine in relation to literature or art, and numerous other categories.

All of these are relevant to medicine in some degree, and all are competing for attention. Space is limited. How to use that space to the best advantage? The editor must decide according to his own lights. He must determine what is best according to his own scheme of values. I will indicate my own solution to the problem.

Types of Reviews

I have established three types of notice: a mere listing; a short comment called "In Brief," printed unsigned; and a full-length signed review. Each of these categories has its own criteria of selection.

Since there is not enough space even to list all the books we receive, I make a selection on the following grounds: I eliminate those written primarily for paramedical personnel, even though these form an increasing proportion of the total number of books we receive. Similarly, I eliminate books intended for laymen, rather than members of the medical profession. I discard those that are lurid and sensational, and those that are highly technical or show extreme specialization, or that would interest only a minuscule percentage of our readership. With these criteria I eliminate 30 to 40 percent of all the books that come to us. The remainder I subdivide into various categories and list under appropriate headings: surgery, public health, pediatrics, sociology, and the like. The listing provides title, author, and simple bibliographic data. A reader, by scanning the headings, can at least get acquainted with new publications, and often the title alone may arouse his interest. Of the books listed, about a quarter to a third will be subsequently reviewed.

In the category called "In Brief," I present short unsigned notices, limited to 60 to 70 words, that provide a terse indication of the contents, with a few words of evaluation. The full-length signed reviews generally range from 200 to 450 words. As editor I indicate the approximate length of the review I want, and I insist that, with allowable variation, the reviewer adhere to those limits. Some reviewers will ramble to two or three times the length requested. If I asked for 350 words and receive 700, I chop the report down to size or return it to the writer for shortening. Excessive length for one book means that some other book will be shortchanged. Since I insist on limiting the length, I can print 250 to 350 signed reviews a year, and 50 to 100 "In Briefs," in addition to the listings.

The Reviewer's Task

If we ask, What constitutes a good review? we find that the word "good" has a certain ambiguity. It may refer to qualities of the book or it may refer to the qualities of the review itself, considered as an essay in its own right. If we were to hear someone say that a novel had a good review, we would understand that the book was being commended; but if we say, "That is a good review," we mean that the review itself embodies excellence as a particular art form, regardless of the book being discussed. If we think of a review as an art form, comparable to a short story, or a sonnet, or a one-act play, we can ask, What qualities make it good?

First of all, a good review must indicate the contents of the book in appropriate detail. How much is that? The answer relates directly to the nature and construction of the review. The reader must gain the sense of what the book is all about and learn, for example, whether it accords with his interests. But, more important, there must be sufficient detail to create a basis for evaluation, and this will obviously vary according to the nature of the comments. The actual amount of detail expounded in a review will depend on its balance and artistry. Too much detail is bad, and the estimation of too much or too little is a matter of critical judgment.

In addition to indicating the contents, the review must provide some sort of critical evaluation, implicitly or explicitly. The reviewer must have certain standards in mind and must make clear whether the book meets those standards. The reviewer's artistry relates, in part, to the way in which he conveys to the reader this evaluation and the standards on which it depends.

As a third requirement, the book review must be well written. It must constitute a complete essay and, even if short, should be a finished literary production. Although as a literary form the essay is not as popular as formerly, nevertheless the book review (and the editorial) still can—and should—maintain a proud tradition.

It is easier, perhaps, to approach the subject by indicating first some qualities of the bad review. One fault is a lack of structure. The reviewer may ramble from one topic to another in a rather incoherent fashion. Or he may direct his remarks not to the reading public but to the author, and discuss, out of context, specific points understood only by those who have already read the book. Or the reviewer may parade his erudition and enumerate obscure references that the author neglected to cite, or detail many small errors of fact. Still more objectionable is any tendency to discuss not the book the author wrote but the book the reviewer would have written if only he had thought to do so. All these indicate an excessive egotism in the reviewer. I do not want to imply that a review should be impersonal; on the contrary, it is essential for the reviewer to project himself, but not to excess.

On the other hand, many reviews, avoiding these faults, emerge as utterly vapid expositions. They indicate the contents of the book by listing the chapter headings, sometimes with a few sentences that tell what each chapter is all about. Then the reviewer may point to occasional errors, bestow a few words of praise or blame, talk about the illustrations, references, index, and binding, and end by mentioning the audience for which the book might be most appropriate. Reviews of this character are easy to write. They do fulfill the minimum requirements but are singularly lacking in imagination and flair and do nothing to establish the review as an art form.

The good book review requires imaginative insight. Instead of describing the contents by dutiful enumeration, it provides, somehow, a synthesis, a grasp of the whole. The discerning reviewer appreciates the problems that the author faced and can evaluate the merit of proferred solutions. He also may be able to relate the book to significant cultural or professional trends. And he should do this in a relatively brief compass, with careful precise writing. All this requires hard work and considerable experience. The good review is perceptive, informative, and skillfully written, a complete essay, short but with high literary quality. It needs thought and effort, structure and order, revision and polishing, all of which, however, lie concealed in the finished product. Furthermore, the good review has a sparkle that emphasizes the individuality of the writer. Good reviews are rare.

The book reviewer is a critic. As generally used, *criticism* is an unpleasant word, for it implies fault-finding. We often hear the comment, "I have no criticism to offer," which means, "I have no fault to find." But this association of criticism with fault-finding, although common, is only a special sense of the word, an extension of its original meaning, namely, the process of judging, or indicating the merits and demerits, or, in a terminology that I prefer, of *evaluating.* The critic has to do primarily with values and standards and the deviations therefrom.

The critic helps to create and maintain standards of excellence and exert an educative function. He has his own concepts of good and bad, which he tries to make more widely known and more generally accepted. Perhaps the critic is comparable to the salesman: the latter tries to persuade the public to buy a given product, while the critic tries to sell ideas and values, to persuade the public to see things in his his way. We need look only at such towering figures as Samuel Johnson in the eighteenth century and Macaulay, Matthew Arnold, and Ruskin in the nineteenth, to realize how much critics have influenced the whole course of literature. Today, in the field of literature, critics are still influential, although perhaps less so than in earlier times.

A medical reviewer or critic must first of all have professional competence, a command of the field in question. He must be able to speak with authority adequate to command respect for his opinions. Obviously, the type and degree of competence will vary with the subject matter: a book on medical sociology will require an expertise different from that demanded by a book on the repair of hernias. The editor has the responsibility of seeking as reviewers only those who are competent in their field. But professional expertise, although indispensable, is not enough. A reviewer should also be able to write clearly and well. Professional competence and writing skill will suffice for the average reviewer, but a further quality is necessary to make him a true critic: he should have a background, perspective, and insight, and an appreciation of historical development and the factors relevant thereto. These qualities, difficult to describe precisely, are essential if his judgments are to possess any depth and exert influence.

In the matter of book reviews the reader has certain legitimate needs. He wants to know what new books are being published and which ones might coincide with his interests; and whether those in the latter category would be valuable to him. Should he buy them, or expect his hospital or university library to buy them? To a large extent the publishers try to answer these questions by sending him catalogues, brochures, fliers, and other advertising material, or exhibiting the books at scientific meetings. An advertisement does much that a reviewer does—tells the prospective purchaser what the book is all about, often in considerable detail; tells him about the author or authors; and provides subtle (or not so subtle) evaluations. When a reliable publisher, with an established reputation for excellence, brings out a new book by an author renowned in his field, the book will indeed find a warm reception without reviews.

However, more often than not the author is not well known; the publisher may have a less than sterling reputation; or the publisher may be a new one and have trouble getting contracts with established authorities. Furthermore, often the books are of marginal interest to a given reader, who may be justifiably hesitant about accepting publishers' blurbs at face value. In all these situations the reader may look to his medical journal to provide guidance.

The evaluations that a publisher sends out are of two sorts—those written by the publicity department or one of the editors; and quotations from the opinions of others, usually from published reviews or prepublication readers. But an ad, even when it quotes liberally from the opinions of others, is automatically suspect. The public recognizes a built-in bias, inherent in all advertising. And when a publisher's ad gives a short comment from a book review, how does the reader know what else the review might have said, that the publishers saw fit to omit? Of course, a great deal depends on the reputation of the publisher. But even the best of publishers may bring out a dud, and even an unsavory publisher may once in a while bring out a worthwhile book.

In theory a book review provides an unbiased report, the evaluation of an expert who has no ulterior motives in offering praise or blame. Many reviews are unfavorable; an advertisement never is. But while a reviewer will certainly be free of commercial bias, he may—and probably does—have intellectual bias. To show the extent to which this can go, I recommend the correspondence column of the *Times Literary Supplement*. This journal has a corps of outstanding reviewers who nevertheless arouse vigorous comment from readers who feel that the reviewer was unfair and from authors who feel themselves misinterpreted or maligned. *Quis custodiet custodes ipsos?* Who will watch the watchers? Ultimately, the reading public.   171

And who will influence the reading public? The critics. A splendid circularity.

The Editor's Critical Function

To some extent the book review editor watches both the critics and the public and acts as a sort of overseer. First of all, he chooses the books that are to be reviewed, and thereby can put into effect his own scheme of values. He soon learns that he cannot please everybody and that he should not even try. He must please himself, that is, he must stick to his own values. At all times some people will be displeased, but that would occur no matter what course he took. And if his values are reasonably in tune with the times, and if he is conscientious, he will persuade more and more to his way of thinking. In my editorial decisions I give less prominence to medical texts of technical nature and more to writings that have a broad general appeal—biology, sociology, economics, history of medicine—works that enlarge the physician's perspective, even though they do not advance his technical skill or knowledge. I might choose for review a work on ecology, and pass by a monograph on, say, biochemistry. I have not heard complaints about this policy.

In selecting books for review I must make a preliminary judgment whether the book is "worthwhile." I put this word in quotes because such a judgment, even though absolutely essential, is necessarily fallible. Sometimes I learn, too late, that a book I had passed over was really quite significant. On the other hand, since good reviewers are scarce, I do not want to waste good reviewers on bad books. Sometimes I consider a book worthy of review and send it out, only to be told it is not worth the space. Then the question arises whether a bad book should receive any notice at all, and whether an unfavorable review should be printed when space is short. For the most part, I think not, but I also believe that to print only favorable reviews would be a great mistake. I believe an occasional highly unfavorable review, if not too long, is quite salutary for maintaining standards.

The editor chooses the reviewers, and herein lies a nest of problems. The first responsibility is to pick someone who is competent. This is ordinarily determined by a man's published work, or by the excellence of his training, or by recommendation. The next task is to persuade the competent person to write. Here I often find great resistance. Some authorities accept an occasional book review as an obligation to the medical community. Others refuse to be bothered— the book does not appeal to them, a worthwhile review would be too time-consuming, and they are too busy. Then, to make the editor's life more difficult, many times a well-qualified man agrees to review a book but never finishes it and ignores follow-up letters and phone calls.

But all these are really only minor annoyances. More significant are the questions: Does he have writing skill? Does he have a critical sense and a set of values with which I can feel sympathetic? Obviously, an editor does not try to persuade a reviewer what to say— such a course would be intolerable. But the editor, who is responsible for policy, need not again call on a man whose values are discordant with his own.

The editor must have his own sense of values. Then, by his selection of books and reviewers, he can himself exert a critical function. And in time he may be able to insinuate his own values into the medicoliterary world.

11

Setting Up a Course in
Medical Writing

In the past dozen years or so I have received many requests for advice on starting courses in medical writing. My correspondents, for the most part, were faculty members in medical schools who, presumably discouraged about the writing ability of their students or residents, wanted to bring about an improvement. Through correspondence I made numerous suggestions, and this final chapter is a suitable place to draw these suggestions together.

My own experience has extended over many different kinds of courses. Sometimes I have been asked to give a single lecture; other times to conduct a class for half a day; or to give a course (or participate in one) lasting a whole day or a day and a half. I have conducted classes that lasted for six weeks, but for various extrinsic reasons the series had to be terminated after two years. Most satisfactory, in my opinion, were those courses that lasted one week.

Assume a willing student—medical undergraduate, resident, or physician in practice—wants to improve his writing abilities. What can he expect from a relatively brief course or workshop? And what should an instructor have in mind as reasonable objectives?

Goals and Format of a Short Course

Writing skill I would compare to physical activity. The man who has avoided exercise for a dozen years or so, whose muscles are flabby, and who puffs on climbing a single flight of stairs, cannot get back into shape by going to a gymnasium for a day or two of instruction and then returning to his former way of life. In a few sessions at a gymnasium he may indeed learn some suitable exercises, but these he must continue for a long while until he has established new physical habits. He might have a desire to run five miles without collapsing, but he would first need to improve his overall physical state. In a few hours of instruction he would not learn how to run five miles, but he might learn some sensible exercises which, if followed conscientiously, would generally improve his physical powers and in time permit him to achieve his ultimate goal—provided he had a high enough motivation and enough persistence.

If appropriate exercise is the important first step in physical fitness, the analogue for writing skill is sensitivity training. The student must become constantly aware that much of what he reads is bad. He must learn to recognize it as bad and must acquire enough 175 analytic skill to realize why it is bad. Or, conversely, he must learn to appreciate good writing and the properties that make it so. In a brief workshop or a short course a student can certainly make a good start toward developing sensitivity of this kind. His further progress will depend on how earnestly he reflects, how steadfastly he maintains his interest.

The mere recognition of good or bad is only the first step. Once the student has acquired a degree of sensitivity, or recognition and appreciation, he must learn to transform bad writing into something better. He must develop his skills. Again we can draw a parallel between a physical activity such as tennis or golf or swimming and the process of writing. A well-designed workshop, even if relatively brief, can give a student a good start. His further progress will depend on himself.

A brief course in writing should try to impart sensitivity, provide some sort of analytic framework for making evaluations, and indicate various techniques for achieving improvement. Obviously there is no single "best" way to reach these goals. No two instructors will conduct their classes in the same way. I can make some suggestions, drawn from my own experience.

For me the optimal number of students per course—or section of a course—will range from eight to fourteen. If less than eight constitute a class, there will not be enough give-and-take discussion. If more than twelve or fourteen, there will not be opportunity for everyone to have his say. The class sessions must be conducted on an informal basis. The instructor is not an authority figure, but rather a resource person, more experienced than the other members but nevertheless only *primus inter pares.* He is not necessarily right; he must not try to impose his views. He is merely "one of the boys" who happens to be more experienced. He has, however, two positive functions: to provide the material and lay out the program, and to act as a moderator and keep order in what should be spirited discussion.

Although the overall subject matter is exposition rather than "creative" writing, the material for discussion should come from all classes of writing. Sensitivity training involves awareness of the values of words and of the different ways of putting words together. All categories of prose are grist to this mill—advertisements, news reports, fiction, essays, editorials, scientific papers, textbooks, legal documents, government regulations, all would be suitable for analysis. At the beginning of a course the leader should provide abundant short passages, representing good and bad writing, comparable to those offered in Chapter 2. These provide material for preliminary analysis and discussion. But the students must also bring in examples—drawn from magazines, newspapers, books, documents—that seem to them either good or bad. This will encourage students to be more critical in their reading, and that is the first goal of the workshop.

Discussion should focus first on the problems, What is wrong with this or that passage? and then, How can you improve it? After a short while the students will become quite adept at spotting faults and correcting them, and exercises of this type will then have diminishing returns.

A second major category of exercise is style analysis. The leader should bring in samples drawn from writers noted for their distinctive styles—offering them in pairs that differ widely—for example, Tom Wolf and Henry James, Ring Lardner and Edith Wharton, Hilaire Belloc and the "President's Page" from the average County Medical Society Bulletin. *How* do the individual samples differ one from the other? *How* does the author achieve his affect (or fail to achieve it)? Exercises of this type demand careful analysis. Initially the student may need special guidance, and the leader must encourage all the students to participate. But through persistence they will gain increasing insight into the value of words and the force of particular constructions.

In discussions of style, occasional students will want to spend valuable class time in small grammatical points such as the status of a split infinitive, the "proper" usage of a semicolon rather than a comma, or the validity of this or that abbreviation. The leader should firmly keep away from what he considers minutiae that do not affect the main features of the exercise—acquiring a feeling for words and for alternative ways of saying something.

I suggest a threefold division of a course. First, the feeling for words and constructions; this the student will acquire if he carefully studies examples, drawn from all types of writing. Second, problems of revision, directed toward the work of others; and third, problems of revision, directed toward one's own work.

Many special assignments dealing with the work of others can promote facility with language. For instance, the leader may take a few pages of a rather technical paper and ask the members of the class to rewrite it for an unsophisticated audience at a high school level—i.e., popularize the message. Or perhaps condense the article to half its length, or write a summary in no more than 100 words. Many exercises might aim at promoting virtuosity, like finger exercises in piano technique. Thus, the leader might provide a page of a suitably chosen text, with the injunction to eliminate all the passive constructions—not some, but all, without exception. This may involve a considerable degree of circumlocution and the use of constructions vastly different from the original. But as an exercise, even though purely academic, it is remarkably effective in teaching the nuances of language. The leader may run a contest in what I call syllable-paring—reducing the total number of syllables by finding shorter synonyms wherever possible. The student will have the task of seeking one-syllable words for those initially having two, two for three, three for four or more, with the absolute prohibition of any word having more than three syllables (the suffixes *-es* and *-ed* are not considered to add to the syllable count). This helps to induce a splendid facility with words.

As a matter of good teaching technique, the leader should perform the same exercises as the students and offer his own efforts for class discussion. He will often find that one or another student comes up with a version superior to his own. A leader should not hold out stubbornly for his own version. Humility is the key to successful group leadership. Sometimes the leader is *secundus inter pares.*

Working with the writings of others is a means of acquiring skill in language, and to achieve this the ingenuity of a leader will suggest many more exercises than I have noted here. But all this is subordinate to the ultimate goal of the entire course—improving the students' own writing.

Here, however, we face problems of a different kind. When dealing with the writing of outsiders, class discussion can take place in a relatively calm atmosphere, and ego-bruising will remain at a relatively low level. But people tend to be much more defensive when something they themselves have written comes up for discussion, and tensions may develop in classroom give-and-take. The leader must decide whether discussion of an individual's work should be a matter of private conference or classroom free-for-all. My own decisions always incline to public discussion of selected essays, but this involves certain logistic problems, centering on what the student wants to have analyzed.

What will the student bring to class to serve as a basis of discussion? Often the announcement of a course in writing leads applicants to expect some disinterested criticism of work they have already written. Those who want to attend usually have a manuscript of their own with which they are not happy. They want suggestions for improvement. Most courses that last more than a single day will ask applicants to submit samples of their writing. In various programs in which I have taken part, even though the directions explicitly asked for samples of no more than one page, most members sent in entire manuscripts that might range from five to as much as 20 pages.

Such an offering is worse than useless. The leaders become intensely irritated, and no one profits. Long manuscripts have no value for instructional purposes, cannot serve as the basis for class discussion. What the students had really wanted was not training in the principles of sound exposition but a hand-tailored critique of their own papers. For this we cannot blame them, but we can blame them for being utterly unrealistic. They want private tutoring at the expense of the class.

Those who want to take a course in writing should, in my opinion, send in samples written specifically for the occasion and strictly limited to one typescript page. These examples should be essays complete in themselves and not merely fragments of larger works. The applicants who sent in entire previously written manuscripts were trying to save effort. However, anyone who wants to save effort should not take a course in writing, whose very essence is effort, and more effort, and still more effort. For a single-page essay there are innumerable subjects: an editorial or expression of opinion on some topic of interest, professional or general; com- 179 ments about a book; a set of directions for some procedure; the plan of a research project; or, if nothing else comes to mind, "Why I want to take this course." The essay should be expository prose, it should be an entity in itself that requires organization, and it should be brief enough so that it can serve in its entirety as a subject for class discussion.

This essay constitutes a baseline. During the course the student should rewrite it, using the lessons learned while studying the writings of others. Presumably he will have sharpened his critical faculties through exercises carried out in the course. Can he apply this heightened insight to improve his own work? During the course selected essays, in both the "before" and "after" form, should be photocopied and distributed for class discussion. The authorship should not be disclosed, although after an hour's discussion anonymity will usually no longer obtain. Free discussion of this type can be enormously stimulating.

Of course, if there are a dozen members of the class, time will not permit class discussion of everyone's work. The leader will choose those papers he thinks will be most profitable for the group. All students, however, will have the assignment of revising their initial essays. This is the focus of the entire course—not to produce literary critics but to enable the students to write more clearly and effectively. Studying the works of others, noting mistakes and infelicities and ways to correct them, provide the student with a technique. The point of the course is to have him apply these techniques to his own work.

The best way to do this is to have the student rewrite what he has already written. And then repeat the process. Two revisions are better than one. The short essay that constituted the admission card for the course would serve as the first effort. The number of further assignments of this type would depend on the length of the course. If it lasts a week, one or even two additional essays can be requested. If the course is less concentrated, with meetings once a week stretched over many months, the schedule could be altered accordingly, with more essays or more revisions.

Improvement in writing skill comes from thoughtful revision and rewriting. Various technical aspects and hints for overcoming difficulties I have already touched on in Chapters 3 and 5. Here I would mention only the need for special exercises, each relatively brief, each subjected to careful scrutiny at intervals. I do not want to imply that revision should be necessarily tied in with class discussion. Discussion merely acts as an intensifier, making vivid the lessons to be learned, and permitting all to benefit from a single example. For maximum benefit there should also be individual conferences between student and the leader. The major emphasis should be on revision.

The student must gain a perspective that only a lapse of time will bestow. With time and experience he will appreciate the importance of small changes—deleting a word here, substituting a word there, adding a few words at an appropriate place, transposing a word or phrase—these can make an enormous difference. And the student should be prepared for what we may call a reverse twist—that which he introduced as an improvement at one time he may want to cancel when he reexamines the writing at a later date.

Those who offer courses on better writing will often include the problems of manuscript preparation—proper organization; conforming to the time-honored straitjacket of Introduction, Materials and Methods, Results, Discussion, and Summary; punctuation; references; illustrations and their display; and comparable topics. All these are important in their own sphere but in my opinion should not be included in any course in writing. After all, what is the course trying to do? We should think again of the comparison with physical activity, and liken a course in writing to a few hours spent in a gymnasium. The besetting sin of many courses is the attempt to do too much, to cover all phases of writing, manuscript preparation, editing, and production, in a few relatively short periods. We would not think highly of a physical culture instructor who wanted his students to perform on the parallel bars, run a five-mile race and lift 100-pound weights, all after a few periods of instruction.

One source of difficulty lies in the prevalence of lectures as a means of instruction. Many courses on writing depend chiefly on lectures, with perhaps a few short periods of small-group discussion. I find no merit in such programs. Lecturing is quite easy for the teacher, and since the problems of writing, editing, and production are so numerous, these can occupy many hours of talk. But in a lecture the listener tends to be passive. To be sure, a lecture may help to transfer some aspects of knowledge from one who knows to one who wants to know, but acquiring information is not the same as developing skills. One learns to play tennis not by hearing a lecture on the subject but through active effort and much sweat. Learning to write requires just as much effort, although the perspiration may be of the nonphysical kind.

Although formal lectures accomplish little, brief informal presentations of data and of problems can profitably initiate any group discussion. Such presentations would serve as orientation, however, not as primary instruction.

Many courses in writing flounder because the goals are not clear in the minds of the sponsors. As already noted, a course should have three parts: the development of sensitivity—gaining an awareness of faults and learning techniques for correction; applying to the writings of others the sensitivity thus gained; and then applying it to one's own work. If these represent acceptable goals, how concentrated should a course be? The answer depends on certain extrinsic considerations.

Group Study and Self-Instruction

Many large medical conventions may offer, as an auxiliary part of their programs, a workshop in writing, lasting a day, a day and a half, or even two days. Such an offering can enroll a relatively large number of people, drawn from many different parts of the country. But the results are rarely satisfactory. The greater the enrollment, the greater the reliance on lectures and the fewer the opportunities for small-group discussion and practical exercises. Any benefits that might accrue under these less-than-optimal circumstances are too easily dissipated. There is no chance for reinforcement.

An alternative is a special course of limited enrollment, lasting a week or ten days, devoted entirely to problems of writing. However, relatively few professional people can afford that much time for this specific subject. A third possibility would involve a group, well localized geographically, that could meet once a week for several months and thus spread the exercises over a long time span. Here, however, there might be difficulty in finding, in a narrow locale, enough people with deep interest and the persistence to carry the project through in the face of unavoidable distractions. Only an unusual group would have the strong motivation and the necessary stubborn persistence.

A few years ago there were impressive advertisements offering instruction in creative writing, supervised by "famous writers" and all carried out by mail. While I am rather skeptical of the degree to which famous writers actively participated, I am confident that a great deal can be learned through instruction-at-a-distance. Such a mode of instruction intensifies the efforts that the student himself must make and to some extent guides him.

The learning process depends on deliberate active effort: Teaching consists of helping the student to learn. However, formal teaching is by no means essential. A young child, with no effort on the part of the parents, learns various skills through watching an older child. The child is self-taught. The learning process takes place easily, and solely through inner impulsion. Adults require vastly more effort, deliberate and persistent, and for the most part need the outside stimuli that we call teaching. Instruction by mail can help, even though probably inferior to personal contact.

Adults, however, can achieve competence in a particular field, unaided by the organized and formalized stimuli we call teaching. Such persons we then call "autodidacts," a term more popular in Europe than in this country.

A program of complete self-study is entirely feasible despite the absence of group interaction. Such a program requires ingenuity and great self-discipline. Yet if these qualities are sufficiently strong, and adequate initial talent exists, the autodidact will have an assurance, independence, and individuality that may well be envied by those who emerge from a classroom.

I hope that this book may help all autodidacts achieve their goals and that it will serve participants in somewhat formalized class exercises.

Index

185